UNPLUGGED PARENTING

UNPLUGGED PARENTING:

How to Raise Happy, Healthy
Children in the Digital Age

Dr Elizabeth Kilbey
with Heather Bishop

First published in Great Britain in 2017
By HEADLINE HOME
An imprint of HEADLINE PUBLISHING GROUP

Cataloguing in Publication Data is available from the British Library

Trade Paperback ISBN 978 1 4722 5048 3
eISBN 978 1 4722 5047 6

Typeset in Dante MT Std by Palimpsest Book Production Ltd, Falkirk, Stirlingshire

Printed and bound by CPI Group (UK) Ltd, Croydon CR0 4YY

Headline's policy is to use papers that are natural, renewable and recyclable
products and made from wood grown in sustainable forests. The logging
and manufacturing processes are expected to conform to the environmental
regulations of the country of origin.

HEADLINE PUBLISHING GROUP
An Hachette UK Company
Carmelite House
50 Victoria Embankment
London EC4Y 0DZ

www.headline.co.uk
www.hachette.co.uk

To Rita and Roy

Contents

All the case studies in this book are based on real families I have dealt with in my clinical work. However, to protect patient confidentiality, names and some identifying details have been changed.

Introduction

Why do we need to unplug?

There is a ticking time bomb in our children's lives. It's in our schools, nurseries, homes, in our bedrooms, in our living rooms and under our roofs, easily accessible 24 hours a day. It is causing arguments in families and it is affecting our children's brains, their behaviour, their weight and their development. It is changing the way our kids play, how they socialise, and how they spend their time. That time bomb is screen use – and the frightening thing is that most parents feel powerless to stop it, or change it.

As a leading clinical child psychologist, I have helped hundreds of children and their families, and I have seen that screen time worries parents more than any other issue. It is the one worry that unites nearly every single parent I see, and it is the most talked about topic in modern parenting.

Most parents are gravely concerned about the amount of time their kids are spending in front of a screen, yet they feel confused about what the true risks are and what they should be doing to manage this problem. At the end of 2016 an Ofcom report revealed that the internet has replaced TV for the first time ever as the most popular pastime for UK children.[1] Recent research shows that almost half of British parents fear their children are addicted to screens – with a staggering 47% worrying that their children spend far too much time 'plugged in'.[2] Combine this with the fact that there are no official government guidelines on managing screen time and it is no wonder that parents feel lost, confused and fearful.

The worrying truth is that the effects of long-term screen use on children still aren't fully understood – but, as I will illustrate, the evidence we have so far is deeply troubling. I have seen some of those effects first-hand in the children I have worked with. It's not unusual for me to see adolescents who spend so much time online that they are unable to sleep, or perform at school – such as the fourteen-year-old boy who was online for 19 hours a day, who refused to go to school, and who wouldn't leave his bedroom. This might sound so alarmingly extreme that it's easy to assume this is a rare and highly unusual event, but unfortunately it is not. Increasingly, I come across young children who are so absorbed in their online time that they will wet themselves rather than leave their screen to go to the toilet. One child I treated had taken to defecating in takeaway containers in his bedroom rather than spend any time away from his online game, such was his obsession with remaining on his device. When I started my career twenty years ago, I had never encountered anything like this: it was unheard of. But twenty years ago the internet was not commonplace: it was not in all our homes and workplaces like it is now. I could not have predicted the extent to which it would change my work – and the lives of the children I work with.

Why is there a problem?

> Too much screen time for kids 'could cause long-term brain damage'.[3]

> Spending too much time online is 'causing mental illness in children'. [4]

Every day parents are reading headlines such as these. Their message is loud and clear: too much screen time is damaging our kids. We don't know how to respond to this, and it's all too easy to think the problem doesn't apply to your children or you. Much of this problem, I believe, is due to the fact that we learn how to parent from our childhood experiences and from growing up with our own parents. We have a model, or code, of how to parent that comes from our own experiences of being parented: in psychological terms, this is called an 'internal working model'. We have a set of rules or ways of doing things, 'a moral map' that's passed down through families and which helps guide us. For example, what children should eat, when they eat, where they sleep. When it comes to parenting your children, your decisions about these things will come largely from what you did as a child. Did you eat as a family, or did the children eat first and parents later? What time did you go to bed? Did you co-sleep as a child, and is that commonplace in your family? However, our children are the first generation of 'digital natives', meaning they have been born into, and are being brought up in, a world where their experiences, their learning and their home life are heavily influenced by digital media, in a way that ours was not. From the day a child is born, digital devices are part of their world. Whereas up until a few years ago a baby's birth would have been announced in the newspaper or with a written card posted to friends and relatives, now news of their arrival is passed on via email, text or by being announced on Facebook and Instagram. Chances are, a smartphone or a tablet will be in the delivery room with the parents to immediately capture their baby's first moments on a screen. One study showed that babies are now so used to digital devices in their lives that over a third of them learnt how to use a smartphone before they could even walk or talk.[5]

As parents, historically we simply have no experience to relate to that is in any way comparable to the way our children live today. We don't have any lived experience of screen time on which we can draw, which is why it's so troubling. Managing devices in your child's life is not part of our internal model of parenting because it wasn't part of our lived experience growing up. We don't have any knowledge to draw on or guide us based on what our parents would have done with us. This is the first time we have been faced with a parenting challenge with no guidance at all from the previous generation. At the same time, we are exposed to a raft of conflicting information about how to manage screen time, from pseudo-science to scare stories in the media. We worry about the impact the digital world is having on our children. Will screen use lead to our children growing up too fast? Are our children losing their innocence at a young age because of what the digital world is exposing them to? Is screen time going to affect our children's attention span and ability to concentrate, or cause hyperactivity or violent behaviour? Will they become addicted to a digital device? Or are we holding our children back by not embracing the digital world, so placing them at a disadvantage educationally and socially?

Let's compare the digital world to sugar, another global health crisis affecting our children. Most parents wouldn't dream of handing a biscuit tin to their child every day and letting them help themselves to as many biscuits as they wanted, whenever they wanted. But many parents don't think twice about giving their child unregulated access to a digital device such as a tablet or a smartphone. Are the effects of screen time really less damaging to our children than the effects of sugar?

I was driven to write this book because the effect of the digital world on children increasingly concerns me. I've witnessed a worrying change in our children: play has changed from a physical, creative experience using toys and imagination to something that involves sitting down in front of a screen, alone, for hours at a time. Parents can see for themselves the all-consuming, absorbing lure of the online world for their child. Their children frequently don't listen to them any more; they refuse to do homework; getting them off their screen is an endless battle; and the first question children often ask when they wake up in the morning is 'Can I go on the tablet?'

However, I would also strongly argue that the current situation is not all bad. Far from it. I can see the huge number of positives of the digital world. For example, it has changed the way I deliver my therapy – from showing children YouTube clips and prescribing information websites to parents, to sharing children's interests in current games as a way to communicate with them. I've also experienced the impact of the digital world personally as well as professionally. As a mother of three children, I am familiar with the issues involved in parenting children in the digital age. I can see how rapidly things have changed between my eldest child (who got a phone at eleven and a tablet at fourteen) to my youngest (who has had access to an tablet since the age of four and could look up things on YouTube before he could even write). I've struggled with my own worries about how to integrate and exclude technology from my family life, and now I want to pass that knowledge on to other parents. I want parents to know about the dangers the online world poses, and to feel more confident in the choices they make regarding their children when it comes to screen time.

Is banning screen time for children the answer?

I don't believe that banning screens is the solution. My career in psychology (and working with children) has taught me the importance of being realistic and practical. Banning screens is impossible in today's switched-on society, especially as a lot of schoolwork and homework is now done electronically, and some schools even give pupils their own tablets to work on at school and at home. The digital world is here to stay. Parents have no choice about whether to get on board with technology or not: instead, we have to work with it.

I think part of the problem is that any official advice from 'experts' is quite detached from what children and families are actually doing in reality. It is a bit like sticking to a diet – it all sounds good on paper, but putting it into practice is very different. I spend the majority of my working life going into people's homes. When I walk in the door, I can guarantee an adult will shout: 'Turn that tablet/phone/computer/console off!'

So if a ban is not the answer, what is? I believe the most important thing for parents to do is to create, and teach, good digital habits. They should talk to their children about the online world to ensure they have a healthy relationship with screens. I want parents to begin to understand that unregulated access to the internet can lead to all sorts of difficulties. Rules around screen time should be in place for your children – just as you have rules for everything else, such as bedtimes, homework and behaviour. Parents need to start early and create good digital habits for their children *before* it becomes a problem. Many parents I meet have only become aware that screen time is a problem in their family when their child hits adolescence. But the bad news is that if you leave it until the teenage years to try to manage your child's screen time, it may well be too late. It's much harder to take back control once your child is already spending the majority of their time in front of a screen, and their patterns of behaviour have been established. This is particularly the case in the hormonal minefield of adolescence.

The very best time to create good online habits and stop screen time from becoming an obsession is in 'latency-age' (or primary school) children.

What is latency, and why is it so important?

'Latency' (which covers the period from the age of four to about eleven) is arguably one of the most neglected, yet most important, stages in a child's development. (By 'neglected', I mean that there are few books on this stage of child development, and less information and support for parents. There are many more books on babies, toddlers and teenagers.)

We know from a psychological perspective that this period is crucial to a child's development – but, alarmingly, it is also a time when many parents take their eye off the ball.

In the first few years of a child's life, they reach a lot of developmental milestones. They learn to roll over, sit up, crawl, walk, babble and talk. During these demanding and exhausting years, there is a huge amount of support and advice available to parents (such as health visitors, and endless books on everything from sleeping to weaning to tantrums and potty training). However, by the time children start school, life for parents, at least on the surface, seems easier. Children have, by this point, achieved the majority of their developmental milestones, particularly motor and social milestones. They can walk, talk and dress themselves. They are gaining independence. This is the period of calm after the toddler tantrums – and before the teenage terrors. It's the part of childhood in which the rate and acquisition of skills appears to slow down. Things become more stable and settled.

Although things may appear dormant, this is actually a crucial stage in a child's development. It is 'dormant' in the way that a garden is dormant in winter. There might not be much to see above ground, but hugely important things are going on beneath the surface.

This is a time when the developing brain is highly 'plastic' (neuroscientists use this term to describe the ability of the brain to organise or reorganise itself to meet changing demands), and is moulded and shaped by the experiences it has. Thanks to advances in neuroscience, we now understand that young brains are plastic and are constantly making new neural pathways in response to a child's environment and experiences, so we have a far greater responsibility to think about what experiences they have than we ever realised before. And it is also the period when the lure of screen time really takes hold. If, at this crucial stage, children are constantly plugged into a device, then what impact is this having on their developing brains? And, equally importantly, if they are constantly plugged in they are missing opportunities to develop the all-important social and emotional skills they will need for life.

Much of the emotional work of latency is to store up energy for what lies ahead – namely, the rapid changes children will go through in adolescence. We know that primary-age children are open and impressionable. They are like sponges, taking in information that will enable them to successfully navigate the world around them and preparing themselves for the storm of adolescence. Latency is also when children start to develop their own identities and interests and begin to form their own social relationships. As they enter the wider social world of school and begin to meet new people and have experiences outside their family units, they start to develop their own ideas and interests.

This is also a very important stage for parents. This is a time for them to enjoy their child's growing independence and to establish the stable, secure relationship and connection with their child they will need to help them weather puberty.

The rapidly changing digital world is upsetting the developmental calm and stability that latency requires. Screen time has created a new battleground within families. In homes around the world, parents and children are clashing about how much time they are spending online. This is the crucial window for parents to teach children good digital habits about online safety and how to use their judgement, before children become adolescents and they have a great deal more independence.

Latency is all about socialisation – making friends and beginning to develop some autonomy and independence in preparation for growing up. However, if our children are constantly glued to a screen, then we are at risk of raising a generation of socially isolated children with poor social skills. By the time children become teenagers, over-reliance on (or obsession with) screen time threatens to impact on every part of their life, from their ability to make friends to their academic achievement. Just one example of this is a Cambridge University study that recorded the activities of 800 fourteen-year-olds, then subsequently analysed their GCSE results. Children who spent just one extra hour per day on screens saw a fall in their GCSE grades equivalent to two grades.[6]

In this book I will:

✪ help parents understand the dangers of too much screen time on a latency-age child

✪ help teach parents what dangers to look out for at each stage of their child's development

✪ give them the tools to stop screen time becoming a problem, so they can bring up healthy, well-balanced children

✪ help them take back control of their child's online behaviour and create a safe family environment where screens can be used positively.

Chapter 1

The effects of screen time on a latency-age child

How screen time could be affecting your child's development, including their physical development, play, learning, concentration and social skills

I have been a clinical child psychologist for over a decade and have worked in the NHS for nearly twenty years. In that time, I have seen so many changes in the range and types of problems that children experience. Perhaps most surprising of all is a decrease in the age at which some of these problems can occur. Alongside this, I have witnessed – both professionally and personally – the rapid rise of technology. I strongly believe there may be a link between some of the difficulties young children are having and their use of technology at a younger and younger age. As I said in the Introduction, I am most concerned for children in the middle age-range of childhood, which is often called the latency phase.

In the past decade, the amount of time that British children spend online has more than doubled. In 2005, eight- to fifteen-year-olds were online for 6.2 hours per week. By 2015, the average time spent online had increased to 15 hours. Children start to go online younger too – in 2014, according to Ofcom, 47% of three- to seven-year-olds used tablets with internet access. By 2015, this had risen to 61%.[1]

It is clear that more children than ever are using screens and they are on the internet for much longer than before. I work with children and families around a range of day-to-day issues and challenges they face, right up to serious mental health difficulties. I specialise in behaviour management and helping children with a range of issues, such as anxiety and low mood. I help families to understand why their children are behaving the way they are, and what they can do to support them. I also work with teenagers and young people, and have gained an invaluable insight into the life of today's teenagers.

Most parents I work with worry about the effect that spending time online has on their children. Whether it is the temper tantrums children throw when parents try to remove a device from them, or children who are bored, uninterested and unwilling to do anything that doesn't involve a screen, these parents intuitively know that screens are having a negative effect on their offspring. Many professionals are clearly troubled too. On Christmas Day 2016, the *Guardian* published a public letter written by forty clinicians, academics and authors about their concerns that a screen-based lifestyle is harming children's health.[2] In it they expressed how they believed children's health and well-being are being undermined by 'the decline of outdoor play' and 'increasingly screen-based lifestyles'. They stated: 'If children are to develop the self-regulation and emotional resilience required to thrive in the modern technological culture, they need unhurried engagement with caring adults and plenty of self-directed outdoor play, especially during their early years (0–7)' and they called for national guidelines on screen-based technology for children up to the age of twelve.

It is interesting to note that some of the world's most famous 'tech gurus' are very low-tech – or even no-tech – when it comes to their own offspring. When a journalist once commented to the late Steve Jobs that his children must love the latest iPad, his response was: 'They haven't used it. We limit how much technology our kids use at home.' Many executives in Silicon Valley reportedly send their children to Waldorf Steiner schools, which exclude screen time before the age of twelve.

But what effect are screens having on our children – and their childhood? As a starting point, I have listed here the developmental milestones a latency-age child should be meeting – and when they should be meeting them.

Aged four	
Social/emotional issues	**Language/communication**
Most children: • enjoy doing new things • play 'mums and dads' • are more and more creative when playing make-believe • would rather play with other children than by themselves • coopcrate with other children • cannot often tell what's real and what's pretend. They talk about what they like and what they are interested in	Children will: • know some basic rules of grammar, such as correctly using 'he' and 'shc' • sing a song or recite a poem from memory • tell short stories • be able to say their first and last names

Aged four (contd)	
Cognitive (learning, thinking, problem-solving) issues	Movement/physical development
Children should be able to: • name some colours and some numbers • understand the idea of counting • start to understand time • remember parts of a story • understand the idea of 'the same' and 'different' • draw a person with two to four body parts • use scissors • start to copy some capital letters • play simple board or card games • tell you what they think is going to happen next in a book	Children should be able to: • hop, and stand on one foot for up to two seconds • catch a bounced ball most of the time • pour, cut with supervision, and mash their own food

Age five	
Social/emotional issues	**Language/communication**
Most children: • want to please friends • want to be like their friends • like to follow rules • like to sing, dance and act • are aware of their sex • can tell what's real and what's make believe • show more independence • are sometimes demanding, and sometimes very cooperative	Children will: • speak clearly • tell a simple story using full sentences • use the future tense (for example, 'Grandma will be here tomorrow') • say their name and address
Cognitive (learning, thinking, problem solving) issues	**Movement/physical development**
Children should be able to: • count ten or more things • draw a person with at least six body parts • print some letters or numbers • copy a triangle and other geometric shapes • know about things that are used every day in their world, like money and food	Children should be able to: • stand on one foot for ten seconds or longer • hop – may be able to skip • do a forward roll • use a fork and spoon, and sometimes a knife • use the toilet on their own • swing on a swing, and climb

Age six	
Emotional/social issues	**Language/communication**
Most children:	Children will:
• have fears, such as a fear of monsters or large animals	• produce most sounds accurately, though may still have difficulty articulating certain letters properly
• want their parents to play with them. Parents are their main source of companionship and affection. A gradual shift begins, though, to fulfilling more of these needs with friends and other people they admire, such as teachers	• have fluent speech by now: indeed, it may seem as though they never stop chattering
• play in ways that include a lot of fantasy and imagination	• be able to produce speech that is generally intelligible, and sentences should be mostly grammatical
• like to be the oldest, and feel as if they are taking care of a younger child	• be able to give their full name and know their age, birthday and where they live
• usually like to play with friends of the same sex	• understand common opposites, such as big/little, heavy/light and under/over
• start to understand the feelings of others, with the encouragement of parents and other caregivers. But they are still most focused on themselves	• develop increasingly descriptive and detailed language
	• recognise when words are unfamiliar, and may ask what they mean
	• be able to read at least ten easy words, such as 'cat' and 'hat', and read simple books

Age six (contd)	
Emotional/social issues	**Language/communication**
• begin to develop a sense of humour. They may like simple jokes and funny books and rhymes • be able to tell a coherent story about people and objects in a picture or predict when one event will follow another, such as going to the park after school	• copy short words accurately, and may be able to write some words unaided • do increasingly detailed and sophisticated drawings and paintings • know how many fingers and toes they have • often be able to count up to 100, repeat three numbers backwards, and understand the concepts of 'half' and 'whole'
Cognitive (learning, thinking, problem-solving) issues	**Movement/physical development**
Children should be able to: • tell you their age • count to, and understand the concept of, ten. For example, they can count ten sweets • learn to express themselves well through words • start learning to write • start to grasp the concept of time	Children should: • be starting to lose baby teeth • have an increasing sense of body awareness and balance • have better coordination – they can hop, skip, jump, walk more steadily on low walls or beams, catch a ball in their hands without clasping it to their chest, and may learn to ride a bike

Age six (contd)

Movement/physical development

- be able to distinguish left and right
- have more developed fine motor skills, so they have more control over a pen and write and draw more accurately
- be able to dress themselves and tie and untie shoelaces
- be involved in more focused play, such as doing more complicated jigsaws or building more complex structures with bricks

Age seven

Social/emotional issues	Language/communication
Most children:	Children:
<ul><li>are becoming increasingly independent</li><li>are keen to be liked and accepted by friends</li><li>are able to take turns and play cooperatively</li><li>will form strong bonds, can be very supportive of other children, and may have a best friend</li><li>tend to play with others of the same sex</li></ul>	<ul><li>should have mastered most speech sounds and be talking fluently</li><li>will be able to describe common objects and explain what they are used for, and follow instructions with three separate components</li><li>should be curious about the workings of their environment and like to hold long conversations, tell jokes or copy accents</li></ul>

Age seven (contd)	
Social/emotional issues	**Language/communication**
• can play board games and understand rules – though they may want to adapt them or add some of their own (and will often cheat if they can) • lose some of the carefree attitudes they may have shown in earlier years, as an increasing appreciation of reality sets in and they start to think more about the future. This is often the age at which they stop believing in Santa • may start to worry about not being liked	• should understand, and use, opposing terms such as same/different or start/finish, and word analogies such as cat/drink, car/boat/plane or walk/swim/fly • start to read by themselves, and appreciate age-appropriate books • should be able to write quite a few words, copy more complex shapes such as a diamond, and understand some symbols, such as & or = • will now produce more detailed drawings (for example, a picture of a house may include paths, garden and sky) • should be able to tell the time on an analogue clock to within a quarter of an hour

Age seven (contd)
Movement/physical development
Children should: • have well-developed coordination. They may seem full of energy and be keen to show off physical abilities • be able to throw, catch and kick a ball, ride a two-wheeled bike, and perform manoeuvres such as handstands • be eager to practise and improve their skills: they may develop an interest in activities such as climbing, swimming, dancing or football • be able to dress themselves easily, write clearly, and use implements such as scissors well. They often enjoy model making, painting, drawing and craft work

Age eight	
Social/emotional issues	**Language/communication**
Most children: • enjoy being around their friends. The opinions of their friends become increasingly important – and peer pressure may become an issue • gain a sense of security from being involved in regular group activities such as Cubs • are more likely to follow rules they help create	Children: • have well-developed speech and use correct grammar most of the time • are interested in reading books. For some children, it is a favourite activity • are still working on spelling and grammar in their written work. This aspect of language development is not as advanced as speech

Age eight (contd)
Social/emotional issues

- have rapidly changing emotions. Angry outbursts are common
- can be critical of others, especially their parents. They may seem dramatic and sometimes rude
- can be impatient. They like immediate gratification, and find it hard to wait for things they want
- are interested in money. Some children may become obsessed with saving, and plan to earn and spend money

Cognitive (learning, thinking, problem-solving) issues	Movement/physical development
Children should be able to: - count in 2s (2, 4, 6, 8 and so on) and 5s (5, 10, 15, 20 and so on) - know what day of the week it is. They do not usually know the full date and year - read simple sentences - complete simple single-digit addition and subtraction problems (such as 1 + 8, 7 + 5, 6 − 2, 4 − 3) - tell the difference between right and left	Children should be able to: - tie their shoelaces - draw a diamond shape - draw a person with sixteen features - show increasing skill in hobbies, sports and active play

Age eight (contd)
Cognitive (learning, thinking, problem-solving) issues
They tend to: • have a black-and-white perspective much of the time. Things are either great or awful, ugly or beautiful, right or wrong • focus on one trait or idea at a time, which makes it hard for them to understand complex issues

Age nine	
Social/emotional issues	**Language/communication**
Most children: • are becoming more emotionally mature and better able to handle their frustrations and deal with conflicts • may experience mood swings and be prone to outbursts of anger, but they begin to cope with, and recover from, these emotions better • are past the stage of believing there are 'monsters under the bed' and are able to let go of unlikely fears	Children will: • have speech that is almost at an adult level, and they can understand and use an extensive vocabulary and complex sentence constructions • be able to think more independently, plan ahead better, think critically, and have improved decision-making and organisational skills • have a longer attention span, and will usually be intensely curious about the world and how things work

Age nine (contd)	
Social/emotional issues	**Language/communication**
• may feel more pressurised by the need to succeed, and may experience anxiety, especially about school performance • are increasingly independent, but still need support and security from parents	• often spend prolonged periods absorbed in activities that interest them, and may read in depth, want to learn more about a topic, and share their thoughts and opinions • understand that objects can be classified into categories and may enjoy collecting things • be able to perform mathematical operations such as addition and subtraction with multiple digits, and be able to understand and use fractions and organise data • be able to give detailed accounts of events and topics and complete more complicated school projects, though some may now struggle more with increasingly challenging academic work

Age nine (contd)

Language/communication

- be keen to belong and to fit in, increasingly defining themselves through their peers and having a strong sense of group identity and loyalty
- be enthusiastic about doing more things away from parents and home, such as having sleepovers with friends
- often have other adult role models, such as teachers or sports coaches, but they are increasingly influenced by children of the same age and may be susceptible to peer pressure
- be aware of social rules, and be capable of behaving appropriately in most situations
- have learnt to be more careful with their belongings
- Will often have a strong sense of right and wrong. Many children become increasingly socially conscious by this age, and start to express opinions about fairness, helping others and making the world a better place

Movement/physical development

- Physical growth may start to diverge between boys and girls, as puberty can start any time from now
- Both sexes continue to increase steadily in height and weight, and gain coordination and strength
- Girls experience an earlier spurt in height, and may now be taller and heavier than boys of the same age

Age ten	
Social/emotional issues	**Language/communication**
Most children: • enjoy being with their friends. They often have a best friend of the same sex • continue to enjoy team and group activities • insist they are not interested in children of the opposite sex. But they may show off, tease, or act silly as a way of getting attention from or interacting with them • like – and listen to – their parents. Some children, though, will start to show irritation with or lack of respect for adults who are in charge	Children will: • enjoy reading. They may seek out magazines and books on subjects of special interest • converse easily with people of all ages • have speech patterns that are nearly at an adult level

Age ten (contd)	
Cognitive (learning, thinking, problem-solving) issues	**Movement/physical development**
Children should: • know the complete date (day of the week, day of the month, month and year) • name the months of the year in order • be able to read and understand a paragraph of complex sentences • be reading chapter books • be more skilled in addition and subtraction, and building skills in multiplication, division and fractions • have learnt to write in cursive script • be able to write simple stories	Children should: • have developed control of their large and small muscles. They are able to enjoy activities that use these skills, such as basketball, dancing and football • have developed endurance. Many can run, ride a bike, and enjoy activities that require a degree of physical fitness • continue to improve their fine motor skills, such as those needed for clearer handwriting and detailed artwork

How screens are affecting our children

Below I have listed the main areas in which I have seen a rapid change since the growing use of digital devices in latency-age children.

Physical problems

It is common for me to see children who are not reaching their developmental milestones because of their reliance on digital devices. At the lower end of latency (age four/five), I see more and more children whose motor skills are impaired because they spend so much time sitting down. This is reflected in things like their walking and their gait – for example, they may still go up and down stairs step by step, like toddlers do, because they have not yet learnt to alternate their feet. These children have not developed their gross motor skills during their early years, so they don't have the agility or the coordination to hop, skip, jump and balance. Another major concern is all that sitting, staring at a screen, has meant they have not developed their core strength. This means their posture is poor, and they cannot sit because their core does not hold them still. So when these children go into school and have to sit still in a more formal setting, they are unable to. Instead, they fidget and wriggle. This leads to them moving around, distracting themselves and other children. Ironically, an early childhood spent indoors on a digital device does not equip children for the task of sitting still at school, whereas children who have run around and are active have good muscle control and are able to sit still better. I go into a lot of schools now and see many of the children using wobble cushions (which are flat, round cushions with a bumpy surface). They give the child an unstable platform on which to sit, thus stimulating their core muscles to work to help them develop postural control and enabling them to be able to physically sit still.

There is an array of physical problems that have been linked to too much screen time in early childhood. Chiropractors have reportedly seen an increase in a condition they have described as 'text neck' where teenager's spines are abnormally bent because of time spent bending over tablets and phones. One study found that children who play video games for more than an hour a day have a higher chance of having wrist and finger pain.[3] Interestingly, the lead author of this research was an eleven-year-old boy who noticed his fingers ached, and he wondered if the pain had been caused by playing games on a Wii. With the help of his rheumatologist father and researchers from New York University, he handed out questionnaires to 171 of his schoolmates who were aged seven to twelve. About 80% of them reported playing with game consoles or hand-held devices. The study found that each additional hour of use per day increased the likelihood of experiencing pain by 50%. Younger children were also more likely to have wrist pain than older children, and researchers put this down to the fact that their muscles and tendons are still developing. However, because the children were not examined, the cause of the pain (or whether it was likely to cause long-term damage) remains unknown. Excessive use of the internet has also been linked to poor vision in children.[4]

How children play and why screens can affect that

Too much screen time is preventing latency-age children from learning how to play. At the very start of latency, children are at the parallel-play stage – meaning they play alongside other children, but not often with them. This is because they find the social aspects of playing, such as sharing and negotiating, quite difficult. They find it hard to demonstrate prosocial behaviour and abide by the polite rules of engagement, as they are still governed very much by their impulses and can be competitive. To put it simply, at this age it is more about snatching than sharing. By the time children start school, they are beginning to get much more interested in playing with others, and they understand that having a playmate can be much more fun than playing alone. In this early part of latency there is a big emphasis on imaginative play and make-believe. A lot of play at this age is about making and creating things, and most children love boxes, sticks, dens and mud. It is also about imitative play and learning about the real world, so popular toys include cars and vehicles, toy kitchens and pushchairs, rockets and doctor's kits. Latency-age children need to experience the real world – this is hugely important for their development. If they spend most of their time on a device, they are missing out on lots of real-life learning.

Play skills are like all other developmental skills, in that they need to be practised and rehearsed in order for them to improve. Children learn to play and they get better at playing the more they do, so it's vital they have the opportunity to develop their creative, imaginative and social play

Play is crucial for a latency-age child's development. In fact, there is a vast array of evidence to support the fact that play is the predominant medium through which children learn in their early years. If children have not developed the full array of play skills because of too much screen time then there is a very real risk that they are going to find starting school difficult. The Early Years curriculum is a programme of learning centred on play and self-directed exploration, so children can go to different tables within the classroom and choose which activities they want to do. Children who have had a lot of screen time tend to flit from one activity to the other and often struggle to focus on the activities being offered – because these require more advanced play skills than they possess. These children do not have the imaginative and creative play skills that children of their age should have. Many of the preschool children I come across in my clinical work who have had large amounts of screen time tend to be stuck in the basic, simplistic play of a toddler, which is all about cause and effect. They will spend their time bashing two things together or filling things up and emptying them out rather than taking part in more imaginative play or role-play that is symbolic of life around them (such as playing mummies and daddies, police officers or fire fighters). This can be incredibly frustrating for a child, because they are struggling to engage with the activities being offered, and means they are more likely to be disruptive within the classroom and find it hard to make friends. I am not convinced that any aspect of using a digital device helps a latency-age child to develop and practise the play skills they are going to need in the classroom or the real world.

Focus and concentration

The internet is fast-paced. You are encouraged to click quickly from one link to another, never fully digesting a whole piece of text or giving yourself time to get bored. A survey of Canadian media consumption by Microsoft found that the digital age has left humans with such a short attention span that even a goldfish can hold a thought for longer.

Researchers surveyed 2,000 participants and studied the brain activity of 112 others. They asked participants to do tests such as responding to pictures, spotting differences and answering questions about numbers and letters. They found that the average human attention span (across all ages and genders) has fallen from twelve seconds in 2000, around the time the online revolution began, to eight seconds.

This means we now have a shorter attention span than a goldfish (nine seconds). Such is our impatience that half of us won't wait for more than three seconds for a website to load. According to research by digital performance measurement firm Dynatrace, just a half-second difference in page-load times can make a 10% difference in sales for an online retailer.

The online world encourages everything to be delivered at a rapid pace and in bite-size chunks of information: news is condensed into 140-word tweets and we text and send emojis rather than have conversations.

The problem here is that children's developing brains become used to this rapid pace of information. They become used to instant gratification and reward – if they perform a task correctly in an app or complete a level in a game, then they win a virtual certificate or a sticker flashes up. Learning in the real world is much slower and a lot less instantly rewarding. Children who spend a lot of time on digital devices might learn to become good at multitasking, but is this at the risk of them being able to focus and concentrate? If children cannot focus, then this is a huge problem for their future education.

Researchers have shown a link between screen time and a shortened attention span. One study found that children who exceeded two hours per day of screen time were 1.5 to 2 times more likely to have attention problems in the classroom.[5] The same study also showed that these children showed less ability to exert self-control, and tended to be more impulsive.

Although some researchers have pointed to a possible link between children's screen time and attention deficit hyperactivity disorder

(ADHD), I remain sceptical. ADHD is largely a genetic condition and it could be, as I will explore in Chapter 8, that children with ADHD are more attracted to screen time than neurotypical children, therefore they spend more time online and playing games – although their screen time hasn't caused their ADHD.

At work, I now frequently meet with children who find it almost impossible to sit through a 50-minute session, even when they are provided with materials to keep them entertained. I always have my travelling kit of child-friendly toys with me (such as puppets, cars and colouring pens) that children can use when I am talking to their parents. But now I am increasingly seeing children who can't occupy themselves and who request a mobile phone or tablet to play on while we are talking.

Learning and education

Many parents try to justify their child's screen time by telling me they largely use their device to access educational apps and games, so therefore it must be OK. However, I always question that: in my experience, children generally access a *lot* of apps and games, with the educational kind only being one small part of that. One US survey found that less than half the time children aged between two and ten spent in front of screens was actually spent on 'educational' material.[6]

My concern about educational screen time is that children are being mentally spoon-fed; they are not being allowed to work things out for themselves. They might know all their numbers, letters and phonics by the time they start school, which is great, but a latency-age child needs a breadth of learning, and numbers and letters are only one small part of it. In order to learn, children need to explore and they need to fail and experience trial and error. Learning through screen time does not allow for any of this. We often describe young children as being like sponges, soaking up everything around them. In the digital world, we are drenching the 'sponge' rather than letting it soak things up.

Screens deliver knowledge to children rather than the child having to seek it out for themselves. Previous generations of children would have had to seek out and find knowledge. Information and learning were hard won. Now children can access information at a click of a button, and the experience of learning is passive and delivered to them. This is a totally different way of gaining knowledge for children today. Information is instant and constantly available, but I don't get the sense that children are building critical researching and evaluation skills. Everything can be Googled: answers are always there, and children don't question or evaluate what they have found. They don't cross-reference sources and deduce knowledge in the way they may have done in the past.

Learning things on a digital device means that kids do not work things out for themselves; they don't have that 'lightbulb moment'. Working something out is an important part of learning. It can often be a struggle, but it teaches children patience and determination to get to the end point.

The most worrying thing, I think, is that we don't yet know what the differences are going to be between previous generations of children, who had to seek their own knowledge and put effort into their learning, and this generation, which is used to being 'given' learning through electronic devices.

In 2016, a video clip about Millennials (the generation born between 1982 and 1994) in the workplace went viral. (Search for it on YouTube!) In it Simon Sinek, a speaker, author and consultant, describes how this generation, which was the first to grow up with technology at their fingertips, is unhappy and dissatisfied. He argues that, largely due to technology, they are impatient and are used to a world of instant gratification. They can have something delivered almost instantaneously: they don't have to wait a week to watch their favourite TV programmes; instead, they can binge-watch them all at once. Sinek argues that some things in life cannot be achieved immediately – like job satisfaction, strong relationships, a skill set, self-confidence and a love of life that require what he describes as a 'slow, meandering, uncomfortable, messy process . . . the overall journey is arduous and long and difficult'. Is this the first glimpse of what can happen to a generation of young people that has had access to screen time from early childhood?

Studies have also shown that screen time can have a negative effect on schoolwork and academic achievement.[7] Researchers measured the changes in the academic performance of non-gamers (boys aged six to nine) four months after receiving their first video game system. These boys showed less involvement in after-school activities, lower reading and writing scores, and more teacher-reported academic problems.

It is worrying that we can already see the effects of too much screen time on the current exam generation – especially as these children did not have the level of technology in latency that our children do now. It's too early for us to know the full impact of a technology-filled latency on the educational outcomes of this younger generation – but, from the limited evidence we have so far, it does not sound good.

A high level of screen time in latency can also affect brain development. From birth to the age of five, a child's brain makes neural connections and develops neural pathways at twice the rate it will do after the age of five. After then, connections are being solidified and the pathways that are used are the ones that are reinforced, while the ones that aren't used dwindle and are 'pruned'. Therefore, if children are doing repetitive activities (i.e. being online all the time), they risk not developing the depth and breadth of neural connections they will need to support their future learning, because the only pathways they are reinforcing are the ones related to digital activities.

Social problems

One of the most disturbing consequences for children who spend significant amounts of time on digital devices is that it dramatically interferes with them developing socially. One of the main features of latency is other children: it is about a child's peer group and developing friendships with them. It's about play dates and having friends round for tea. This is a very familiar experience for many parents, but I am increasingly aware of primary school children who don't need to have friends round to play, they just log on and play with them online. But this is not playing in the full sense of the word. This is screen time.

One of the key developmental functions of latency is that children start to immerse themselves in the world around them, while having the support of their family in the background to scaffold them. However, if they are glued to an Xbox or a tablet, and are not playing or talking to anyone face to face, then they are missing opportunities to learn how to socialise. Digital devices by their very nature are isolating. Parents often try to persuade me that gaming is a sociable activity because children can game online and chat to their friends, but for this age group I remain unconvinced. In adolescence, digital devices become part of a teen's social world, so teenagers text and WhatsApp and FaceTime: devices become a tool to support their social life – an essential accessory in the constant need to stay connected. In latency, I believe they are a barrier to a child's social life. During latency, children start to form close relationships and learn the rules of friendship in a safe way. Much of the time, children socialise under the watchful eye of adults, whether this is the dinner lady at lunch time or other parents when your child is on a play date. At this time, children can always seek the support of adults and they are there to intervene in a dispute or help them negotiate how to play together. Children of this age also know they can retreat to the safety of their family and disengage from socialising for a while. They can disconnect for a while and their friendships will still be there the next day at school. This time spent socialising face to face, with the support and guidance of adults, is essential to help primary school children develop their social and emotional relational skills. I am concerned that time spent on devices prevents children from developing these skills.

One study carried out by the University of California in Los Angeles found that screen time might be inhibiting children's ability to recognise emotions.[8] Researchers discovered that eleven- and twelve-year-olds were much better at reading human emotions after five days of taking a break from technology.

Screens should not replace human interaction. Children need lots of face-to-face time with others to learn social cues such as facial expressions, body language and tone of voice. You don't get any of that from a screen. What I see in my clinics is a generation of children who have missed out on social markers because of spending too much time on digital devices. Schools are having to react to this: many are setting up clubs to teach children the social skills they should have picked up during latency. These are very basic, but incredibly important, skills, such as learning how to have a conversation, learning to how to share, how to take turns and how to see things from another person's perspective. This is alarming, particularly when you see the knock-on effects on a child's education, social development and, ultimately, their emotional well-being.

Case study

Ten-year-old Hannah was referred to me after she began having friendship problems at her primary school. Over the past year she had been increasingly isolated by her friendship group, and had been identified by her school as a vulnerable pupil who might struggle with the transition to secondary school. Her parents were worried, and didn't know why things had changed so much in such a short time for their daughter. Previously she appeared to have had a normal friendship group and was not identified by either her school or her parents as having significant social problems. But Hannah was now very miserable and worried about going to school. She did not have friends to play with at lunchtime and was not enjoying school.

THE SESSIONS

When I talked to Hannah, to find out about some of the difficulties she was having and about her life in general, it soon became clear that one of her main interests was watching episodes of an American teen drama on her tablet. Her school provided pupils with their own tablets, which her parents had not worried about, because it had parental restrictions on it. Hannah could access YouTube to look things up for school, which was where she was watching these dramas – she was watching up to six or seven episodes a day. It became apparent, from talking to Hannah, that she had got so immersed in them that she thought the world was like it was in the programmes. She was really struggling socially because she was finding it hard to talk to the other girls and relate to them, and did not understand the rules of friendship: while the other girls were developing socially, she was absorbed in her tablet, getting left behind. Once you fall behind socially, it is hard to reintegrate yourself.

From having had plenty of friends a year ago, Hannah was now struggling. Other girls were starting to make fun of her obsession with these shows and the fact that they were all she wanted to talk about. She did not know how to join in their games or conversations, and she was making herself more and more unrelatable and marginalised. When she fell out with someone, she did not know how to resolve it. She thought it would work like it did in the shows she watched where the characters, who were of a similar age to her, would fall out, but then they would 'stage an intervention' and everyone would make up again – all in the space of a half-hour episode. Hannah did not understand that the real arena of social relationships was not like this: when she wanted to try to 'intervene' when people had fallen out with each other, they rebuffed her attempts at fixing things, and mocked her for being naive. This left Hannah feeling very upset and confused, so she had retreated further from her peers and even further into the fake reality of her TV shows. She was lonely at lunchtimes and had stopped being invited round to other people's house. The worse things got at school, the more she had retreated to her tablet. Watching these dramas had become her replacement for real-life friendships. She had found an alternative reality to exist in, rather than the real world. As a result, the real world had moved on and left her behind.

THE INTERVENTION

First, I had to help Hannah's parents understand why Hannah had been socially excluded, and how the tablet had contributed to that. Her parents were unaware of what she had been watching on YouTube, and they couldn't understand what the problem was, as there was nothing wrong with these programmes. They weren't violent, they didn't contain bad language, and they were aimed at Hannah's age group. I had to help them understand that, because they were constantly available and so absorbing, Hannah had immersed herself in this fantasy world and it was stopping her engaging with the real world.

I then had to do a lot of social bridge-building with Hannah. I had to help her understand that these TV shows were not real – this was very upsetting for her. We talked a lot about how different things seemed to be in the show compared to how she experienced things in the real world. I helped her understand that real life did not work like TV, and showed her that one of the consequences of spending so much time on her tablet was she had fewer friends in the real world. Once her parents were on board, I got them to start regulating her tablet use so she was limited to watching one episode a day. I worked with her school, who put her in a lunchtime social support group with other children who were struggling in the playground. I tried to build a few more social links for Hannah. We went back to basics and organised play dates with other like-minded girls who lived locally and who were going to go to the same secondary school.

Hannah's parents also worked with the school to try to restrict the way the tablets were used. For example, on the school bus Hannah would watch episodes on her tablet rather than talk to the other girls. The school arranged for the driver to turn off the Wi-Fi so pupils had to talk to each other instead of everyone being on their devices.

OUTCOME

This wasn't a quick fix for Hannah, and it took time for her to develop the enhanced social skills she had missed out on that her classmates had been building for the past year. By the end of Year 6, after lots of play dates and help from her teachers, Hannah had made a couple of secure friendships, which stopped her from feeling so isolated. Hannah's issues also made the school rethink how they were encouraging their pupils to use tablets. It prompted them to think about tablets not only as an educational device, but also as something that could act as a social barrier. The school had good e-safety awareness, but had not focused so much on social developmental work with their Year 5 and 6 girls. They began to address this more actively after my work with Hannah.

Red flags for children aged four to seven

Q Your child finds it hard to do physical tasks their peers seem able to do – such as riding a bike, swimming, climbing, running, PE at school.

Q They struggle with activities involving fine motor skills, such as dressing themselves, doing buttons, laces and zips, or using a knife and fork.

Q They find it difficult to be creative and imaginative during play. They wouldn't know what to do with a cardboard box, or know what to build with blocks.

Q They are not as interested in dressing up or make-believe play as they were previously.

Q They find it hard to make up a game or story using toys such as dolls, figures, cars or vehicles.

Q They have had problems related to starting school, and have found it difficult to settle and make friends.

Q Their behaviour in the classroom has been disruptive.

Red flags for children of all ages

Q There are changes at school – such as teachers commenting that your child isn't concentrating as well and they are less focused than they have been in the past, or school reporting that your child is spending lunchtimes or break times on their own.

- They have stopped talking about friends they used to talk about a lot.

- They seem unhappy, and talk about being 'bored' at school – for older latency children, this often means they are lonely.

- They are no longer interested in activities that they used to enjoy. They are only interested in spending time online.

- They show less interest in tasks they have previously enjoyed which take time and concentration (such as reading, writing, drawing, colouring and Lego).

- They seem easily distracted (e.g. they wander off when you're having a conversation with them. They have a reduced attention span for anything other than screen time).

Some solutions

- Children learn and develop through 'doing' – trying, practising, problem-solving, learning different tasks. Don't replace the 'doing' part of child development with screens. Screen time is often just about watching.

- Make sure digital devices are not the sole focus of their play repertoire.

- Distract and divert. We use this technique a lot for toddlers, but it can also work well with an older child. Distract them by suggesting another task, whether a game or an activity, to move them away from the screen.

- Use the computer in different ways. The internet is full of ideas, so use the screen to lead your child to other activities: for example, ask them to look up videos of how to make a rocket out of a toilet roll, and then encourage them to go off and do it.

➲ Make sure you prioritise real-life play for younger children, rather than virtual play.

➲ Help scaffold social development by providing chances for your child to meet up with other children. Keep mixing up the groups so your child mixes with as wide a range of people as possible

➲ Give your child lots of social opportunities – join clubs, go on group outings, go to parks, playgrounds and soft play centres. These are all places where children can easily meet other children and strike up a conversation.

Chapter 2

Owning a device

Everything you need to know before you buy your child their own digital device, from where to keep it and how to restrict its use.

P eople are often surprised when I tell them that my eight-year-old son has had his own tablet since he was four. I can see the look of relief on their faces, as if they are thinking, 'Well, if a child psychologist has done it, then it must be OK!' Parents often ask me, 'Should I buy my child a tablet? What's the right age to buy my child their own device?' To be honest, the decision to buy my son his own tablet was mainly a practical one, in an attempt to avoid family arguments over devices. However, before he was even allowed to turn it on I made sure we had put clear rules in place about when he could use it, how long he could use it for, and where he could use it. Although my son might have classed the tablet as 'his', I made sure he knew I had control of it.

What is apparent is that over the past few years there has been a steady increase in young children owning their own devices. One in three children in the UK aged between five and fifteen now has their own tablet computer, and one in three eight- to eleven-year-olds has their own smartphone.[1] Even toddlers have their own tech – a recent study found that 38% of two- to five-year-olds own an Android tablet, 32% own an tablet, and 32% have a mobile phone.[2]

So is a latency-age child responsible enough to have their own digital device? Will a sense of ownership actually encourage them to be more responsible, or do parents risk relinquishing all parental control by handing their child their own computer, laptop, smartphone or tablet?

Things to consider before buying a device for your child

When weighing up this decision, there are a few important things that parents need to think about. The majority of latency-age children do not yet have the cognitive skills to be able to delay gratification, which means they have a very limited ability to wait for anything. They also have a poorly developed internal clock, which means they are much less able to understand the passing of time. In relation to screen time, that means if they are left on their own to manage a device, they have almost no concept of time passing, so if no one challenges them or sets any rules then they will be on it continually. They are unable to recognise when they have had enough screen time and they need to put the device down.

Children this age have an underdeveloped self-regulatory system: they are still driven by the 'I want it and I want it now' mindset. This is perfectly illustrated by the famous Stanford 'marshmallow test'. This was a series of studies on delayed gratification carried out in the late 1960s and early 1970s, led by psychologist Walter Mischel, then a professor at Stanford University. Hundreds of children took part in the experiments, the majority of whom were aged four to six. The child was put in a bare room with a marshmallow on the table in front of them, and the researcher told the child he was going to leave the room for 15 minutes. If the child did not eat the marsh-mallow while he was away, then they would be rewarded with a second marshmallow when he returned. If the child decided to eat the first marshmallow before the researcher came back, though, they would not get a second one. Researchers found that two-thirds of the children ate the marshmallow. The older the child, the better they were able to delay gratification – in other words, not eat one marshmallow, but instead wait for a reward of two marshmallows. These experiments show us that younger children find it very hard to wait, even when told to do so by an authority figure. They can't understand the benefits of waiting now in order to have more later. Even if it's in their best interests to wait, they cannot control their impulses. But when they get older, they can use their cognitive skills to override their impulses and they can reason with themselves. Essentially, they can use their cognitions (thoughts) to manage their feelings (urges). This is an advanced, but vital, psychological skill that children need as they move through childhood into adolescence.

Latency-age children are also unable to juggle competing demands – and their developing brains can't multitask yet. So give them a screen without any interruptions – and the likelihood is they will allow themselves to sit there all morning and not get dressed or brush their teeth or tidy their room or do any of the other tasks you might have asked them to do. This is quite a substantial challenge to any family that needs to get of the house in the morning!

My two main reservations about children owning their own device are:

✪ Will it lead to more screen time?

✪ Will it lead to private/unregulated screen time?

Unregulated screen time

Problems can arise when screen time is not being regulated or supervised. That is where parents lose control. A recent study carried out by Internet Matters revealed that many children are literally being left to their own devices online. Alarmingly, 44% of children are browsing the internet, on social media and streaming content from the internet without adult supervision – and nearly half of six-year-olds surf the web alone in their bedroom.[3] The one thing I would stress to parents is that you can't buy your child a screen (whether this is a phone, a tablet, a PC, a laptop or a console), hand it to them, and hope for the best. You need to take responsibility and have rules and a plan in place – as a latency-age child just does not have the skills to manage their own screen time. And why should they, when we haven't put the safeguards in place to stop them? We would not give children unregulated access to anything else that could potentially be harmful if it's not managed (such as sugar), would we?

Owning their own device often gives children the illusion of control. This is particularly true with older children who may have saved up to buy their own device. Their logic is 'I paid for it, therefore this is mine, so I get to decide when to go on it, how long I'm going to be on it for, and where I keep it.' But if they perceive they are in control, that is when problems can begin. Latency-age children have grown out of being tantruming toddlers and are just beginning to test out their own power and autonomy. Therefore it is really important for parents to instil authority – even when the child perceives they are in control. Don't be persuaded by their argument. Don't relinquish your power: this is a vulnerable place to be, and that's true for nearly all aspects of parenting. Imagine if you let your child choose what they ate for dinner each day or what time they went to bed. It's important to acknowledge to the child that, yes, they might have got to choose what they spent their money on, but you are still the parent and you have rules about screen time.

What is your plan?

Before a device even enters your home, you need to have a plan in place. A lot of parents, particularly if it's their first child, can't quite imagine the power a digital device can have and how it can take over if you take your eye off the ball. It's like having a puppy in the house: you are going to have to think about how it's going to work and fit into your home and your family. If a child said they were going to buy themselves a skateboard or a drum kit, then I'm certain most parents would have a plan about how and when they were going to be used. But the majority of parents I meet for whom screen time has become a problem did not start off with a plan in place.

Think about the rules you are going to set around screen time. Is the device allowed at the dinner table? Is it allowed in bedrooms? How long is your child allowed to go on it? When are they allowed to go on it?

There is no right or wrong answer to these questions – parents have got to think about it, decide which rules will work best for their family, and make sure they enforce them. Always stick to what I call the basic parenting principles – be consistent, know when to say no, know when you are going to concede and, above all else, remain in control. One of my most frequent and important pieces of parenting advice is 'pick your battles' – but make sure you win the battles that you pick. You can't get into a battle about everything; if you did, family life would be very stressful indeed. So decide what you are going to be flexible about, but once you decide what you are going to stand firm on, make sure your child knows you will enforce this rule. When it comes to devices and screen time, my overwhelming experience tells me that this is the battle to pick – and parents need to win it.

It's also a good idea for parents to understand how a device works before they hand it over to their child. I have known parents who have bought their children iPods and e-readers and not been aware that their child can access the internet through them. One parent I worked with had a huge shock when they realised their nine-year-old son had been accessing hugely inappropriate websites for months without their knowledge, and had been searching Google and YouTube.

Will owning a device give your child a sense of responsibility?

One of the benefits of a latency-age child owning their own device is that it can help them become more responsible. It certainly did my son good to learn to look after something he valued so much from an early age, and he has never smashed, lost or broken his tablet. Years ago I worked in a school in which teachers were teaching children to take digital pictures to expand on the artwork they were doing. They were Reception children, so the teachers had bought special bump-proof cameras because they thought they would be safer for the children to use as there was less chance of them getting broken. These cameras were big and clunky, and looked like something out of a cartoon. However, they still got bashed about and broken very quickly. So instead the teachers decided to buy proper digital cameras – the ones children were used to seeing adults using. This time, none of them got broken because the children understood that these were valuable items and they had to be careful with them. Children are cognitively capable of understanding when something is valuable and precious. How much care they take of something will relate to how much value they put on it personally – if it's something very precious to them, they will try to take more care of it. If it's got a status to it, then they will attempt to look after it. But you have also got to manage this against their level of dexterity. Latency-age children can be clumsy, and they are more likely to drop and break things than an adult because their fine motor skills are not as good as ours. So it's a good idea to put some protective bumpers or covers around any devices that your children use to prevent damage.

Where should devices be kept?

I'm a firm believer that, wherever possible, children (particularly those aged four to seven) should always be supervised when they are having screen time, and digital devices should be used in shared/communal spaces downstairs rather than alone in bedrooms. But I know that, unlike TVs, digital devices are portable and children can move from room to room, so often it's hard for parents to know where they are and what they are doing. I also know how easy it is to lose track of time when you're a parent. Your child is quiet, so you may think, why disturb them? But, even if you have restrictions on your Wi-Fi or on your children's devices, it is still important for parents to have an idea of what their children are doing online – and what they are looking at. If your child is using the device around you, it also helps you to know how long they have been on it. If they are out of sight, it's much more difficult to monitor that.

Ideally, what parents are trying to do during latency is to instil a culture of openness with their children. If your child is on a device in a shared space then although they are not interacting with you at least there is the opportunity for people to interact with them, whether that's a parent asking what they want for dinner or a sibling wanting to know what they are doing. If a child is on a digital device in their bedroom and they are quiet, then it's less likely that someone will come in and interrupt them – and it is more likely they will be left alone for longer.

Devices compete with social time. Being online/on screen is an insular, solo activity that prevents young children from having real-life interactions. At least if a child is having screen time around other people, it is not so isolating.

What I know from clinical research is that in order to maintain parenting authority you need to maintain engagement with your child. In other words, you need to keep connected. Any time someone in the family is engaging in a solo activity like screen time, then it severs the connection between family members – and this is some-thing to be avoided, particularly for latency-age children.

While babies and toddlers are making huge motor and language leaps, latency-age children are doing more emotional and social processing. They are watching, they are learning, they are developing their personalities and absorbing information about the world. If they have their head buried in a device and you can't see them or know what is going on and what is influencing them, that is a huge concern. Social and emotional development is achieved through observing, copying and relating. Digital devices prevent this.

I also strongly believe that devices should not, wherever possible, be kept or used in bedrooms. Even if a child says their device has been turned off or they are charging it, they should still be removed from the bedroom, especially overnight, as often the temptation to use them is too great. I know many adults who use their phones in bed before they go to sleep or in the night if they wake up. If it's hard for adults to resist the lure of the internet when they should be sleeping, it's definitely too hard for a latency-age child.

Hyperfocus

Have you ever noticed how, when children are on electronic devices, they almost go into a trance-like state? Their eyes are glued to the glow of the screen and however many times you ask them a question or try talking to them, it is as if they can't hear you. This all-consuming focus is known as *hyperfocus*. While it can be useful for some things, it is not great if you can't break it easily.

I think we have all experienced being in a hyperfocused state. The other morning, I dropped my children off at school. I was late and I was rushing to catch a train to get to a meeting. I got to the station with two minutes to spare, and noticed a voicemail on my phone which I decided to listen to in case it was connected to the meeting. Unfortunately, there was a hubbub around me which meant I couldn't hear the message, so I turned my head to one side to aid my focus. When I finally hung up after listening to my voicemail, I realised that my train had been and gone. Even though I was standing on the platform and I was right in front of it, I was so focused on listening to my message that I hadn't noticed the loud noise was my train. That is hyperfocus. It makes you push the background away and focus on one thing. I frequently observe children in this hyper-focused state when on their devices.

I would argue that good attention is not using all of your attention. Good attention is about focusing on a task in hand yet still being alert and aware of what is going on around you. Evolutionarily speaking, we need to keep scanning the environment all the time for survival. In caveman times, hyperfocus would have got you eaten by a sabre-tooth tiger. Or perhaps, in modern-day terms, hyperfocus in a school-age child might get them knocked over while crossing the road as they are using their phone – which is a phenomenon I observe almost weekly. Or it might result in you missing your train and not getting to your meeting on time.

Hyperfocus occurs when something absorbs all your mental energy or attention, such as mental maths or trying to decipher a crackly voicemail message on a noisy train platform, or when something is so stimulating that it grabs all your attention, to the exclusion of everything else – which is exactly what screen time does. Things can be absorbing and engaging, but hyperfocus is a very specific zoned-in state. In it, you are so focused on something that you do not notice anything else going on around you.

Researchers studying ADHD have scanned different parts of the brain to look at hyperfocus in more detail. To put things simply, even when we are in resting mode particular parts of our brains are still working. These are the parts that are involved in scanning the external en-vironment and processing visual and auditory information. These parts of the brain are called the *default mode network*. The default mode network's job is to keep you alert to your environment, so you are aware of potential dangers – to stop you being eaten by a sabre-tooth tiger or being run over. However, when you need to focus on one particular thing or task, those scanning parts of your brain power down and the parts of the brain required for the task activate.

We divert mental resources away from the default mode network to the parts of the brain that are now working on the specific task we have set. We need to switch off the scanning part of our brains so we can focus properly on what we are trying to do, without being constantly distracted by things around us. Research has shown[4] that when people with ADHD are given a task to do, their default mode network doesn't switch off or power down like it is supposed to. They are trying to focus, but they are still scanning the environment around them. That's why they are so inattentive and struggle to direct their attention effectively, because the scanning parts of their brain keep pulling their attention away to things in their environment. When researchers gave people with ADHD something incredibly stimulating to do, though, their default mode network finally switched off – but they went into a hyperfocused state and they couldn't restart their default mode network. This means they effectively got stuck in a 'zoned-in' state and couldn't switch back to tuning in to their environment.

This is the same process that happens when a child zones into a digital device. It is so stimulating that they are only able to focus on the screen, and they are no longer scanning what is going on around them. There are many things about this that trouble me. One of the most serious problems is that a hyperfocused state is a detached state which is not optimum for learning or developing social skills. We all need to be able to respond to the constant stream of information around us. If you are hyperfocused, it can also be very difficult to switch tasks, which is a vital mental skill. Children need to be able to switch focus in order to complete most cognitive tasks.

It is therefore important to avoid too much hyperfocus. I strongly recommend that parents are around when children are using digital devices – it increases the opportunity for environmental stimulation, to interrupt or distract them. Whereas if they are in their bedroom in a hyperfocused state and there are no environmental disruptions they could potentially stay like that for hours. We worry about screen time making our children less focused, but we should also be concerned about too much focus. We want our children to develop flexible concentration skills, so they are able to concentrate and focus when they need to, but they can also switch their concentration when required.

Case study

Four-year-old Evie's parents came to see me as she was coming to the end of her Reception year at school, but she had struggled to settle in and make friends. Her teacher had called her parents in to say that Evie's behaviour and listening skills were poor. Her parents could not understand it; they told me Evie was a bright, capable girl who learnt her numbers, shapes, letters and colours at an early age. They had expected her to thrive at school, as she was academically ahead of her peers, but her teacher had said she was struggling to sustain her focus, and was disrupting the other children in the class.

THE SESSIONS

Evie's parents clearly prioritised education. They were both university graduates, had good jobs, and they wanted their only child to have every advantage in life. They appreciated the value of technology in modern life so they explained to me that they had bought Evie her own tablet when she was two and loaded it up with educational apps. They proudly showed me how she could operate the tablet, and she clearly knew her way around every app. I was impressed by her ability, and she seemed advanced for her age.

But when I went to observe Evie at school, I started to realise she certainly was presenting with difficulties at school – and it was probably the tablet that was causing the problems. Evie knew all her shapes, numbers and colours because of the apps she had been using, but she did not have the social skills or the play skills she needed to thrive at school. Because most of her play had been based around the tablet, she found it difficult to choose, self-organise or self-direct in the classroom. Having been brought up on a diet of fast-moving, noisy, colourful electronic apps, making a block tower or threading beads on to a string was nowhere near as exciting for her. Her peers were happy to play with Duplo or Lego, to play with cars, to play in toy kitchens and push dolls around in pushchairs – but Evie didn't see the value of engaging in those things. Imaginative play was perplexing for her. She couldn't engage with the other children, and she didn't like or understand the toys and activities on offer to her, so she was bored and frustrated. This led to her doing something destructive that would provoke a reaction, causing the adults in the classroom to come and engage with her.

When I spoke to her parents about the tablet, they were shocked that I believed the tablet had led to some of Evie's issues. They thought they were giving their child a huge head start by letting her spend all her time playing on educational apps, and that she would walk into school with an advantage over her peers. I discovered that it had done the opposite. Evie's interest in the tablet had very quickly grown to the point where now, aged four, she spent most of her spare time on it – some days she was clocking up to four hours. While Evie was quiet on the tablet, her parents (like most parents would) had used the time to get on with other tasks.

INTERVENTION

Evie's parents started to realise that it was not a good idea for such a young child to have so much screen time. They had fallen into the trap of thinking the tablet was everything Evie needed to get ahead. However, what a child at the younger end of latency actually needs is interaction and attention. I explained to them how being on a tablet is a solitary activity that was not helping teach Evie social interaction or developing her communication skills or imaginative play – which were what she needed in order to enable her to learn at school. In their attempt to help her, they had made her focus far too narrow.

I suggested that Evie's parents should introduce other activities to broaden the range of things Evie did at home. The focus wasn't on restricting the tablet; it was more on increasing Evie's opportunities to do other things that involved her putting the tablet down. It was as simple as her parents saying 'let's go out into the garden and play' or 'let's go for a walk in the park'. Or doing some baking or drawing with her, playing a game, singing a song together or doing a jigsaw. All these things were going to require a lot more of her parents' time and engagement but, at Evie's age, interaction with others is much more important than interacting with a tablet. Her parents needed to take every opportunity they could to engage and communicate with their daughter.

The tablet could still have a place in Evie's life – if they wanted – but it had to be less available. If she spent more time doing other activities, this would wean her tablet use to a more manageable level. I gave her parents some guidelines to follow, and encouraged them to stop the tablet being their go-to when Evie was at home. I encouraged them to offer her something else to do, ask her to wait, or distract her so she got used to not having the tablet around all the time.

OUTCOME

Evie's parents were very anxious about limiting her time on the tablet. They were worried that she would have tantrums when they tried to cut down her screen time. They realised they had fallen into overusing the tablet because of the demands of their busy lives, so my advice made them stop and try to do things differently. They put a schedule in place about when Evie was allowed on her tablet, and they each freed up some time to play with her during her non-tablet time. What happened surprised them. Evie was so pleased to have more one-to-one time with her parents that she had very few tantrums. There were friction points when she was tired or she wanted her tablet, but it wasn't as bad as they had imagined. In fact, Evie adapted more quickly than her parents had thought she would.

This was not a quick fix, but with a little bit more play and attention from her parents and a lot less time on the tablet, Evie slowly started to settle down in school. By the time she had gone into Year 1 her behaviour had improved, she felt more confident in the classroom, and she had made a couple of friends. As she learnt more about how to play, by doing it with her parents, the tablet became less important to her as she had more interesting things to do with her time.

Red flags

- Parents feel they have no authority over the device; their child controls it. They do not know where it is, when their child is using it, or how long they are on it for.

- There are constant arguments and battles to get their child off the digital device.

- The child keeps their device in their bedroom. If they are not allowed to do this, they will often sneak the device into their room and will frequently be found asleep with it.

- The child is unable to get to sleep, they are constantly tired, or their sleep patterns are disturbed due to use of electronic devices.

Some solutions

⮑ Before giving a child their own device, parents need to ask themselves: does my child understand the concept of time and the value of objects? Are they able to acknowledge that they have had enough screen time, and put down the digital device? If not, they are unable to manage their own device without supervision at this point.

⮑ Regardless of who owns the device, screen use needs to be monitored and regulated. Put firm rules and boundaries in place.

⮑ Ownership can give children the illusion of control – 'This is my tablet, therefore I can do what I want with it.' Parents have to let children know from the start that they are in charge of the device – and that owning the device does not mean the child gets to control it.

Chapter 3

Time online – how much is too much?

How long you should let your child spend online, and the consequences of too much screen time

T he question I am continually asked by parents is: 'how long should my child be online for?'

So, is two hours a day on an Xbox too much for a nine-year-old? How long should a seven-year-old be allowed to play on a tablet? Just how long *is* too long?

So for those parents who accept, as I do, that screen time is now an inevitable part of our children's lives, is there such a thing as an ideal or safe amount of time that a latency-age child should spend in front of a device each day?

If parents are looking for official guidance then unfortunately, in the UK at least, you are out of luck. There are currently no official government recommendations on the amount of screen time children should have. The American Academy of Pediatrics has recently revised its guidelines due to the changing nature of digital technology. It recommends one hour of screen time per day for children aged two to four, and for children aged six and older it simply states parents should 'limit digital media'. It says the emphasis should be on parents to determine the restrictions, monitor the type of digital media their children are using, and place consistent limits on the types of media and the time spent using it. It also says parents should make sure media does not take the place of adequate sleep, physical activity and other behaviours essential to health.

Other countries are more concerned: Taiwanese parents, for instance, are now legally obliged to monitor their children's screen time. The government can levy a £1,000 fine on parents whose children (aged up to eighteen) are using electronic devices for what they deem as 'extended amounts of time' (although, unhelpfully, they don't actually specify what these amounts of time are). The same measures exist for China and South Korea.

From experience, I know that most parents are at best unaware and at worst in denial about how much screen time their children are having. On a Sunday morning it's very common for my son to be up before I am, and he's become very skilled at quietly going on his tablet so he can have uninterrupted time on his device before I get up. Of course, I have no idea how long he has actually been on it, and he certainly doesn't keep track of the time. Many parents I work with will say their child has an hour or so online a day, but when I get them to keep a log of the time they are always very surprised when it adds up to substantially more.

All the evidence points to the fact that latency-age children are clocking up worrying amounts of daily screen time. An Ofcom report on media consumption in the UK estimated that the average three- to four-year-old spends three hours each day in front of a screen. This rises to four hours for five- to seven-year-olds, and 4.5 hours for children aged eight to eleven.[1] A recent survey found that children are spending an average of around 17 hours a week in front of a screen – almost double the 8.8 hours spent playing outside.[2] According to psychologist Dr Aric Sigman, by the age of seven, a child born today in the UK will have spent an entire year of 24 hours a day looking at TV, computer and video game screens. If I told most parents of primary school children that their child had already spent an entire year of their life on screen time, I'm not sure they would believe me. But this is a worldwide problem – a 2010 US study discovered that children between eight and ten spend, on average, an alarming eight hours a day using electronic media outside school – more time than they actually spend in school.[3]

Research has also shown that how long children spend on a screen can affect whether they are likely to become a 'pathological' (or problem) gamer or not. A recent study found that youths who became pathological gamers began by playing for an average of 31 hours per week, whereas those who never became pathological players began playing for an average of 19 hours per week (two to three hours per day).[4] I think this clearly demonstrates how vital it is that we manage the issue of screen time from the moment we introduce devices to our children.

Can too much screen time be harmful?

In a nutshell, yes. Scientists and researchers have only begun to scratch the surface of how screen time affects children, but the initial signs are alarming. Around 2,500 children were studied to assess contributory factors for ADHD. The research showed that, for every hour of TV children watched each day, their risk of developing attention-related problems increased by almost 10%.[5]

The biggest problem is that certain types of screen time can trigger the release of adrenalin. This could be caused by anything exciting, frightening, stressful or stimulating – for instance, watching a football match when you're very invested in the outcome, or playing a computer game in which your character could die. The adrenalin release stems from the perception of a threat, not an actual threat. Triggers include physical threats, fear, excitement, risky behaviours, bright lights, noise and high temperatures – most of which are elements of screen time. So, the excitement your child gets from playing an online game is highly likely to elicit an adrenalin response.

Some of these games will have a 12, 16 or 18 rating: I would strongly advise against giving these to latency-age children. However, lots of popular games aimed at latency-age children, such as Minecraft (which has 7+, 10+ and even 4+ ratings), still have elements of excitement and jeopardy in them.

Latency-age children have vivid imaginations and get caught up in fantasy. They can easily get totally absorbed by the excitement of online games, which activates their body's physiological responses. As parents we wouldn't dream of putting our children on endless rollercoasters and fairground rides several times every day, or letting them run round in a heightened state of excitement for hours on end. We wouldn't do that in the real world, yet we are regularly doing this with our children in the digital world.

Why is too much adrenalin bad for children?

Adrenalin is a hormone the body releases when it is under stress. It provides us with the added energy we need to deal with that stressful situation. One of the adrenal glands' main functions is to instantly ready the body to go into 'fight or flight' mode. Heart rate and blood pressure rapidly increase, our senses are sharpened, and energy is released for immediate use so we can run away from danger.

A problem arises because, every time adrenalin is released, the body also releases cortisol. And cortisol is not helpful in the same way that adrenalin is. Cortisol is known as 'the stress hormone': it can increase your blood pressure and blood sugar as well as lowering your immune system. The prolonged, everyday stress we suspect is caused by excessive screen time can lead to a greater overall production of cortisol. Over time, high levels of cortisol can destroy healthy muscle and bone, impair digestion and stop the body from producing vital hormones.

Cortisol also has an effect on mood, increasing levels of anxiety. Excessive screen time, then, means our children are subjecting their bodies to a chemical attack that is having a detrimental effect on their physical health. It is comparable to the impact of stress on high-pressured city workers – our children are at risk of suffering the physical consequences of 'burnout', but from sitting in front of a digital device.

Cortisol is particularly damaging to latency-age children, the developing brains of whom are not mature or resilient enough to manage the effects it causes. The brain of a latency-age child is still 'plastic'. So while the brain is still laying down neural pathways and acquiring skills – one of the key roles of latency – we really don't want it awash with cortisol and under stress, as this compromises brain development. Beyond damage to the brain, sustained high levels of cortisol have been linked with high blood pressure, diabetes and hormonal imbalance.

What can parents do?

Parents should be mindful of any sort of screen time in which children show a strong over-excitement response, such as an increased heart rate and body temperature, sweating, fidgeting, shouting, restlessness and a faster breathing. These are all signs of increased adrenalin production and should prompt parents to think about how they manage, or limit, exposure to such types of games. This type of response is unlikely to come from watching TV or YouTube – with perhaps the exception of a football match or other sporting event. But is very likely to come from a game (online or offline) that involves fighting or losing 'lives'.

It is important to remember that every child will react differently to different games, apps and online experiences. Parents should follow these guidelines:

- ✪ Observe your child.

- ✪ Consider whether they are showing signs of being in fight/flight mode. Are they agitated/overactive/restless?

- ✪ Be aware of how they behave when they come off their device. Are they aggressive or hyper?

Safe limits

It is very hard to come up with a 'one limit for all' approach, as each child (and how they react to screen time) is different. As a general recommendation I would say children at the lower end of latency (aged four to seven) shouldn't be having more than 60–90 minutes of activating (adrenalin-inducing) screen time per day, and children at the higher end (aged eight to eleven) more than 90–120 minutes. This doesn't include TV – just time spent on digital devices. Ideally this should be broken up into two or three half-hour chunks, which is a reasonable amount of time before and after school.

The benefits of taking a break

Try to avoid children being online for large chunks of time; instead, break it up into smaller sessions. I would suggest half an hour maximum for younger latency-age children, and 45 minutes for older ones. A break is especially important if they are engaging in adrenalin-producing tasks, as being in an agitated state is not good for young bodies, so this should be limited. It is always a good idea to take regular breaks when using a screen, to reduce potentially physically damaging effects, including eye strain (just as we would for PC/screen use at work).

The myth of 'mindful usage'

Some experts claim that, instead of parents controlling and restricting their child's screen time, we should teach our children how to control their own use of screens from an early age, using something they call 'mindful usage'. However, I have to say I am quite sceptical about the possibility of this. Devices and the activities children do on them are, by their nature, all-absorbing. As I said earlier, latency-age children don't have the cognitive skills to be aware of how long they've been on a device and to register that they need to turn it off. Therefore I think that a latency-age child will really struggle to grasp the concept of mindful usage – because of their age and stage of development, I think they still need their parents to do that for them. One study found children get more sleep, do better in school, behave better and see other health benefits when parents limit the amount of time they spend in front of a screen.[6] It also affected body mass index (BMI). Researchers found that if screen time was limited children got more sleep, which also resulted in a lower risk of obesity.

I believe that as children get older, and have greater cognitive skills and ability with self-regulation, then it is certainly possible to work with them to set their schedule and build in some flexibility.

Case study

Six-year-old Sam's parents came to see me because they were concerned about his behaviour at home. He had been having more tantrums lately, and seemed to go into meltdown over the slightest thing. They described how he was quick to get angry and then would shout, throw things and become tearful. Previously, Sam had been a well-behaved boy who enjoyed playing outside. He had a football goal in the garden and liked playing football with his dad. He had recently got a tablet for his birthday, which he enjoyed playing on. He wasn't having problems at school or with friends, and there was nothing else going on in the family to explain the change in his behaviour.

THE SESSIONS

Sam was reluctant to talk to me about his behaviour, although he and his parents both acknowledged that there was a lot more conflict in the house. He was happy to talk about his tablet and his favourite game, Minecraft.

As a starting point, I asked his parents to keep a diary of when these incidents occurred, and to think about what was happening immediately beforehand. They soon noticed that the difficulties occurred around a number of familiar scenarios, including dinnertime, bedtime and doing schoolwork – in particular, they were nearly all times when they had to tell Sam to stop using his tablet.

This also made them notice how often Sam was on his device. When I'd first asked them about it, Sam's parents had estimated that he was on his tablet for about 1.5 hours a day in total. I asked them to start recording how often Sam was actually using his tablet, and they were shocked to realise he was regularly spending up to four hours a day online. He was going on it before school, as soon as he got home from school, and then after dinner/before bed every day, and often even more at weekends.

As we discussed it, they began to realise that Sam was using his tablet instead of doing other things he used to enjoy, such as playing with Lego and playing outside. His parents always had to cajole him to get him off his tablet and to do other activities. The house had become a battleground, with Sam frequently retreating to his room – where he would play with his tablet. His mum, in particular, felt he was much more distant from her and his family, and she could see a change in his mood and emotional well-being. It was clear that Sam was spending way too much time on his tablet, and was obsessed by it – so much so that he carried it everywhere with him.

INTERVENTION

Sam and his parents drew up a schedule together of how much tablet time Sam could have, and when, each day. They picked times that worked for them as a family – e.g. not before school because time was very tight and it was making them late. They adjusted the setting on Sam's tablet so it turned off after a maximum of half an hour's use, and allowed him regular short bursts of the tablet a day – not totalling more than 90 minutes during the week and 120 minutes at the weekend. They also encouraged Sam to do more of the things he used to enjoy, such as play Lego or go outside for a kick-around with his dad. His parents also made a conscious effort to show an interest in what he was doing on his tablet, and discuss the activities and games he was playing.

OUTCOME

Slowly, Sam's behaviour improved. He resisted in the beginning and had a few temper tantrums when he was asked to come off the tablet when the timer went. To start with, his dad had a tendency to give in and allow him extra screen time. However, in a follow-up session I stressed to his parents how important it was to be consistent when they were enforcing the rules, and within a couple of weeks Sam had accepted the new routine.

As Sam's tablet time decreased, it meant he naturally started going back to some of the other activities he used to enjoy. His parents said he seemed much happier and calmer, and there was less conflict in the house. Ultimately, they felt he liked knowing where he stood – and what was and wasn't allowed. These sorts of rules around screen time tend to work with latency-age children, who respond well to boundaries and usually want to please. They generally find rules very reassuring, and they like to stick to them because rules give clarity around what is expected of them and reduce anxiety.

Red flags

Your child is having too much screen time if you see the following changes in their behaviour:

- They are more reactive and prone to outbursts.

- They are easily distressed and tearful and seem either low in mood or on a very short fuse.

- They have temper tantrums, particularly when asked to come off their device.

- They have difficulty sleeping, including finding it hard to fall asleep and stay asleep, or start joining parents in bed when they previously slept OK on their own.

- There is increased sibling conflict and more arguing.

- They answer back to parents and are more rigid/difficult/unco-operative.

Q They have difficulties with friendships – they argue more with friends, or isolate themselves from good friends.

Q They want to stop doing activities they previously enjoyed, and spend all their free time online or on a digital device.

Q They stop playing with previously enjoyed toys, e.g. Lego.

Q They become distressed when they're stopped from going on a device or when they're asked to come offline.

Q They want to spend all their free time on a device. It's the first thing they ask for in the morning/after school.

If online time is interfering with your child's sleep, school work and friendships, and is affecting their behaviour and emotions, then it is time to step in.

Some solutions

◌ With your child's input, set screen rules for the whole family and write them down in a way your child can understand. These can be about anything, including general behaviour (that is, behaviour not relating to screens, such as bedtime, homework, tidying their room) as well as screen time.

◌ Establish a timetable setting out exactly when children can and can't have screen time (e.g. not before breakfast or until they've got their uniform on or after homework). Work out whatever fits best with your family's routine. (You can use visuals/pictures to make the timetable easier for a younger child to understand.)

- Build in some possible exceptions to the schedule, so it is clear and there is no confusion. For example, the child can have an hour on their tablet each day of the weekend, except if there is a family event planned, in which case tablet time may be reduced.

- Limit when and where devices can be used, based on family rules (e.g. not at the dinner table, not in the bedroom). Make sure everyone in the house, including parents, sticks to them as well

- Make sure you monitor your child's use of their device, so you know they are sticking to the rules. (I suspect most parents don't keep that close a track on their child's screen time usage.)

- Try to engage with children when they are online or on a device, and show an interest in what they are doing. Talk to them about what they're doing, playing or watching. If they won't respond during screen time, then talk to them about it afterwards.

- Put a timer on your child's device so they have a clear signal for when they need to come off it. Use kitchen timers or countdown apps so your child knows when the device will be going off.

- Block Wi-Fi from devices when you don't want your children to be online (e.g. before school, after 6pm, etc.).

Chapter 4

The physical problems caused by screen time

The undeniable link between obesity, sleep and the digital world – and how to make sure your child doesn't swap sports for screens

The world is in the grip of a modern obesity epidemic and in the UK we have hit a record high for childhood obesity. According to the latest figures from the National Child Measurement Programme, which measures the height and weight of over a million English schoolchildren every year, over a third of children aged ten and eleven, and over a fifth of children aged four and five, are overweight or obese.[1] Many clinicians agree that screens are one of the major factors that has led to British children being described as 'the least active generation in history'.[2] The statistics make very sobering reading. Half of seven-year-olds are not achieving the recommended goal of 60 minutes of daily physical activity.[3] Traditionally, the amount of physical activity we do tends to decrease as we get older, so there may be serious health implications for a whole generation of children who were so inactive so early on.

Another major study found that one in five children born at the start of the millennium was obese by the age of eleven. The Millennium Cohort Study, which followed 13,000 children born in the UK, showed a sudden surge in obesity between the ages of seven and eleven.

Why is this? For children at the lower end of latency (six and under), there are so many opportunities to run around and be active! Most parents take their children out to the park or soft play, and they willingly run around. As children get older, they start to self-direct a bit more and have more choice over what they do with their free time. This is also when the lure of screens can really kick in, and some children may choose to spend their free time sitting down at a screen rather than running around.

In the course of my work, I encounter more and more children who are failing their Early Years health check (which measures height and weight) and are being described as overweight. However, it appears that problems are starting well before latency – with screen time once again being stated as one of the reasons for this high level of inactivity. A recent report, based on an analysis of government data, found that nine out of ten toddlers are living a sedentary lifestyle because of the increasing amount of time they are spending on gadgets.[4] The report found that just one in ten toddlers are active enough to be healthy, and only 9% of children aged between two and four are meeting UK guidelines (which recommend at least three hours of activity a day). Even the vast majority of preschool children – 84% – do not manage one hour a day on the move.

The consequences of this are alarming. There is plenty of evidence that sitting for hours at a time stores up health problems for the future. A sedentary lifestyle is linked to an increased risk of obesity, cardiovascular disease and stroke. It is also linked to depression, and girls starting puberty much earlier.

A recent study also linked increased screen time to a higher risk of type 2 diabetes. Researchers, who studied over 4,000 pupils aged nine and ten, found that those who spent three hours or more each day in front of a screen were heavier and had more body fat than those who were more active. They also found 'strong associations' between screen time, obesity and risk markers for type 2 diabetes (in particular, insulin resistance).[5]

Physical activity is crucial for children's mental and physical development, and being inactive is the worst health indicator for young children. Low activity levels in young children mean they are not engaging in some of the primary tasks required for healthy development at this stage (such as running, jumping, hopping and throwing). Therefore they are at risk of not developing gross motor skills, balance and coordination.

The journey through latency is vital in order to prepare children for adolescence and adult life: this is the time when children develop the skills they are going to need to navigate this tricky stage of their lives. This includes developing social skills and learning to manage their emotions.

I'm very concerned that the children I see today do not have the basic skills I would have expected to see in children five or ten years ago. Many of the primary schools I work with now ask Reception class children to come into school wearing their PE kit on days they have PE. This is because it takes so long for the children to change into and out of their PE kits! Also, many teachers have explained to me that a large percentage of their Reception children are not able to dress themselves, and obviously teachers do not have time to dress and undress the majority of the class for each PE lesson. This is a relatively new phenomenon – when my daughters were at primary school ten years ago, I had never even heard of this happening.

The link between screen time and obesity

Although no clinician can say that screen time *causes* obesity, the link between the two is undeniable. A recent US study of more than 10,000 kindergarten children found that those who watched more than one hour of TV a day were 52% more likely to be overweight than schoolmates who watched less TV. Children who spent at least an hour each day in front of the television were also 72% more likely to be obese.[6] Researchers have also found that when parents limit screen time it results in a lower risk of obesity for children,[7] and that children who have high amounts of screen time are significantly more likely to become unfit adolescents.[8]

Inactivity and keeping children inside

I believe that one of the reasons for these high rates of inactivity is that society is becoming increasingly risk averse. This is seen most clearly in our attitude towards parenting and the idea of keeping our children 'safe'. In the 1980s and earlier, children would go off with their friends or go out to play and would only come back home when they were hungry. Most parents would not dream of doing that today, because of their perception of risk – abduction, 'stranger danger', car accidents and many more. Nowadays most people have a general sense that it is not safe for children to play outside without constant adult supervision. In response to this perception of risk, parents are now keeping their children at home, mostly inside, so they know they are safe. However, more time inside has translated to more time on screens to help occupy them.

There is a certain irony to this. Parents are keeping their children inside so they are safe and they can keep an eye on them. Yet when children are inside, they are on digital devices, often with limited parental supervision. I would argue that with screen time there also tends to be a more hands-off approach from parents than when children are playing outside, so in actual fact children may be at greater risk of physical and social harm if they are having high levels of screen time.

The importance of allowing children to be bored

Children need to be bored. The ability to self-occupy is a really important skill for a latency-age child to learn. Unfortunately, modern parenting does not allow children to be bored. When I was growing up in the 1970s and 1980s, parenting was much more hands-off. Children were left to make their own entertainment. I certainly don't remember my parents providing me with things to do to keep me entertained. It was my job to find myself things to do and keep myself out of trouble – and that usually involved going out with a group of local friends. Since then, there has been a significant generational shift, and parenting is now a much more active process. Parents are very hands-on and strive to do things with their children, and for them.

On some levels this is very positive because it means we have a much better understanding of our children's needs. But I worry that parents today are becoming what is described as 'helicopter parents', hovering around their offspring, making sure they are happy and stimulated at all times. Parents need to embrace the concept that children can be bored. But what do many parents do if their child says they are bored? They immediately turn to screen time. But a digital device is not the antidote to boredom; it merely distracts a child from being bored. A latency-age child needs to be bored because that is how they are going to develop their problem-solving, creativity, play and social interaction skills. It's also going to teach them to rely on their own resources rather than instantly turn to others for help. A really important part of latency is coming up with your own solutions to situations. A latency-age child should be able to be left to play on their own for a while in the same house as their parents. If they can't, this suggests to me they haven't developed their imaginative and creative play skills enough to occupy themselves. And the way to address that is more boredom, not less!

Making screen time more active

One way of balancing the fact that the majority of screen time is inactive is encouraging your child to try some physically active computer games instead. There are now many consoles on the market that promote physical activity, like the Nintendo Wii, as well as active games like Pokémon Go which encourage players to walk outside and 'catch' Pokémon. There are also dancing games like Just Dance or Dance Central. They still need to be regulated, but studies have shown that playing active video games is similar in intensity to light to moderate walking, skipping and jogging.[9] Researchers also found that active video games led to a small reduction in BMI compared to sedentary video games, which increased it.[10]

Screen time and sleep

Another important link is between screen time and sleep. The amount of sleep children are getting is continuing to decline,[11] and screen time seems to be playing a major part in this. Studies have shown that screen time leads to delayed bedtimes and shorter sleep times.[12] Yet 89% of children play with a tablet or screen close to bedtime, and 92% of parents have admitted they are worried about their children's screen time before bed.[13] Screen time does not just affect the length of sleep a child has, but also appears to have a negative effect on the quality of sleep. One study found that teenagers who played an exciting computer game before bed had significantly shorter periods of REM (rapid eye movement) sleep – the phase of deep sleep.[14] REM sleep occurs at intervals during the night. It is characterised by rapid eye movements, more dreams, and a faster pulse and rate of breathing. During REM sleep, your brain and body are energised: REM sleep is thought to be involved in the process of storing memories, learning, and balancing your mood – although the exact mechanisms are not known. There are five stages of sleep. During the night, we progress from stage one to stage five, which is REM. A lack of REM sleep has been shown to impair the ability to learn complex tasks, and it has also been linked with long-term memory problems, weight problems and migraines. REM sleep is especially important in early childhood, when our brains are rapidly developing. REM sleep makes up a much larger percentage of sleep in babies and children – 20–25% of an adult's total sleep is REM compared to 80% for a newborn baby.

Light is the most powerful cue for shifting or resetting our body clocks. Lower levels of natural light signal to our bodies to release melatonin, the sleep-inducing hormone. The light from digital devices has a higher concentration of blue light than natural light, and blue light affects levels of the sleep-inducing hormone melatonin more than any light of other wavelength. We have low levels of melatonin during the day: our bodies release melatonin for a few hours before bedtime, and levels peak in the middle of the night.

Good-quality sleep is important for health, emotional well-being and learning. If screen time means children are not getting enough good-quality sleep, then this could lead to them feeling sleepy during the day, which is known to influence academic achievement. One study shows us how technology can affect learning and memory.[15] Ten schoolchildren (average age 13.5) played a computer game for 60 minutes on one night of the week, watched television for the same time on another evening, and had one technology-free night. The children had the hour of technology at 6pm, two to three hours before they went to bed, and after a homework session which required them to remember information. They were tested on what they had learnt at the end of the session and then again 24 hours later. The computer game playing led to disrupted sleep patterns (including a 20-minute delay in getting to sleep) and reduced their ability to remember the material they were supposed to have learnt. Watching TV did not have the same effect. The researchers suggested that the reduction in verbal memory may be occurring during the video game play and/or as a result of the disturbed sleep, since sleep also helps to consolidate memory. This is all the more alarming because the results were based on one hour of technology per night (many children have a lot more) that happened several hours before bedtime – so in theory should not have affected sleep.

Screen time and bedtime

My view is that latency-age children should not have digital devices in their bedrooms; as the adage goes: 'out of sight, out of mind'. Parents are less likely to be able to supervise and control when children are using devices, particularly before bedtime or after lights out, if they're not in the same room. Most screen time is visually stimulating and exciting. As I explored in Chapter 3, it can trigger adrenalin and hyperstimulate the nervous system.

Using a screen before bedtime does not allow a child time to wind down and relax. It is hard to fall asleep if you are stimulating the brain and then immediately expecting it to just switch off. I would recommend switching off devices at least half an hour before bedtime – perhaps even an hour before, for younger children.

Good sleep habits for latency-age children

Children aged four/five generally need 10–13 hours of sleep a night, while children aged six to eleven need 9–11 hours. Here are some tips to foster good sleeping habits in your children.

- Try to put them to bed at the same time every night and get them up at the same time, even at the weekend. Children sleep better when they go to bed and wake up at the same time every day. Staying up late during the weekend and then trying to catch up on sleep by sleeping in can mess up a child's sleep schedule for several days. Having the same routine is like setting your child's biological clock.

- Children should only use their bed for sleeping. Lying on a bed and doing other activities (such as watching TV, eating or using a tablet) makes it harder for their brain to associate their bed with sleep.

- Make sure their bedroom is dark, quiet and comfortable, and they are not too hot or cold. Make sure there is nothing too distracting in there and no toys within easy reach of the bed.

✪ Alarm clocks are for waking up. Children who stare at the clock while they're trying to fall asleep should have the clock turned away from them.

✪ Have a bedtime routine. A predictable series of events should lead up to bedtime – such as brushing teeth, having a bath or a shower, putting on pyjamas and reading a story.

✪ Make sure they only do quiet, calm and relaxing activities before bed like reading a book or listening to music. Avoid stimulating activities like screen time (TV or computer) and physical exercise.

✪ Make sure your child has had plenty of exercise during the day. Exercise can help children feel more energetic and awake during the day, have an easier time focusing, and can even help with falling asleep and staying asleep that evening.

Case study

I was initially called in to work with five-year-old Bella's older brother, who had disabilities. When I was working with the family, Bella's mum mentioned she had a few concerns about her daughter. She was worried, as Bella had become increasingly inactive and overweight. As her son needed a lot of extra help, especially in the evenings, and she was a single parent, she had bought Bella a console and a tablet to keep her entertained. Bella kept both in her bedroom. In the past few months teachers had commented on her reluctance to go outside and play at school, and she had recently been identified as being overweight in the school weight check. She was also suffering from regular headaches.

THE SESSIONS

Even though Bella's mum was worried about her daughter she had not made the link between her problems and screen time. The more I talked to her about the devices, the more she realised how much time Bella was spending on them each day, particularly around bedtime. I also suggested that Bella's high levels of online time could be causing eye strain, which could lead to headaches. When I asked about Bella's sleep patterns, her mum said she had trouble falling asleep and would get in and out of bed a lot at bedtime. She was also waking frequently during the night and was unable to get back to sleep. As a result, she found it hard to get out of bed in the mornings and her teacher had commented that she was often tired at school. Because she was tired a lot of the time, her screen time was also affecting her learning and her relationships – she was bad-tempered and argued with her class-mates. It was also making her weight issue worse. Because she'd put on some weight, Bella was more reluctant to go out. Instead, she just wanted to lie on the sofa after school, and was seeking sugary foods to give her an instant hit of energy because she was so tired.

INTERVENTION

As a first step I asked Bella's mum to remove all digital devices from Bella's bedroom. She was still allowed time on her devices, but they had to be turned off at least an hour before she went to bed. Then I worked with Bella's mum to help her implement a bedtime routine with Bella, as she had done when Bella was a toddler. We arranged for a carer to come and help with Bella's brother to free up some important one-to-one time for her and Bella. Bella's mum helped her with a bath and then got her into her pyjamas, and then they had ten minutes in her bedroom, either reading a story or chatting about Bella's day.

OUTCOME

After a few weeks of having a proper bedtime routine in place and no screen time immediately before bed, Bella was able to fall asleep much more quickly – and this helped to reduce her tiredness during the day. She continued to wake during the night and had difficulty in getting back to sleep. I explained to her mum that she had to be consistent in taking Bella back to bed each time she woke up, to help her learn how to resettle herself.

As the sleep situation improved, both Bella and her mum had more energy. Bella's mum also began to see that she could change things for her daughter and she did not feel as overwhelmed. This gave her more confidence and motivation to begin to tackle issues of consistency at home. She felt able to be firmer around the tablet and console and get Bella's use of them down from around three hours a day to a more reasonable hour. This was also helped by her improved relationship with her daughter because they were spending more time together in the evenings. As Bella's mum accepted more outside help with her son, she could find the time to go out to the park or play with her daughter after school. This included walking to school and to the shops. In the long term, it meant Bella would be a lot fitter and would have more energy. Her headaches also eased as her screen time lessened and her sleep improved.

Red flags

🅠 Your child is reluctant to go outside and do any sort of physical activity, such as cycling, running, walking or taking part in team games or sports.

🅠 They have noticeably gained weight and easily get out of breath.

🅠 They have a lack of confidence in their physical appearance.

🅠 They are unable to do similar physical tasks as their peers – such as riding a bike or a scooter, or roller blading. This needs further investigation.

🅠 They find it hard to get to sleep, and you have difficulty waking them up in the morning.

🅠 They seem constantly tired.

🅠 You regularly find they have fallen asleep while on a device.

Some solutions

➲ Work out how much time each day your child spends sitting still or being inactive. Then note down how much of that time they spend in front of a screen. Colour each block of time in so it's easy to see. The results are often a real shock to parents.

➲ Allow your child to be bored. Put a timer on, and say to them if they are still bored in 15 minutes to come back and tell you. You will find that most children will have found something to occupy themselves by then. Don't automatically rely on screens to relieve boredom.

➲ Make being active a family priority.

➲ Make activity compulsory, not optional.

➲ Look at ways to increase your child's physical activity (which will have a knock-on effect on screen time). For example if you walk to school instead of drive, you will have to leave earlier which means there will be no time for any screen time in the mornings.

➲ Encourage your child to enrol in a local sports club or try team games.

➲ Set a 'bedtime' for digital devices – when all devices need to be turned off and charged or put away in one room (ideally, including parents' devices too).

➲ Make sure your child has a proper bedtime routine (which does not include screen time) to wind them down: e.g. a bath or shower, stories or quiet reading.

➲ Do not allow your child to keep digital devices in their bedroom overnight.

Chapter 5

Addiction

How to stop your child from becoming obsessed with screen time – and ways to take away their devices if it gets out of control

Eight year-old Ben had been slumped in a chair, silent, sullen and unwilling to engage with me for most of our session. That is, until I mentioned Minecraft, which his mum had told me he enjoyed playing. Then it was as if someone had flicked a switch: the boy in front of me suddenly came to life. Ben's eyes lit up and he started pacing the room as he animatedly explained about the intricacies of his latest build and how he blew up a zombie with TNT and discovered a rare potion. His mum looked worried as he told me all about this game, which was clearly his number-one passion. Later, she told me:

> It's all he talks about. All he wants to do is play Minecraft, and the more time I give him, the more he wants. It's never enough and it causes so many arguments.

> He sneaks the tablet into his room to play it at night so he's permanently exhausted, and when he's not playing it he's watching YouTube videos of other people playing it. He even dreams about it – I hear him calling out in his sleep.

He should be playing football with his friends like he used to, but he's not interested in that any more. I'm worried he's addicted.

But is Ben's mum right? Can a latency-age child truly become addicted to screen time? Professionals are constantly debating whether Internet Use Disorder (IUD) or Internet Addiction Disorder (IAD) is actually a recognised condition. In May 2013, Internet Use Disorder was added for the first time to the *Diagnostic and Statistical Manual of Mental Health Disorders* (DSM) published by the American Psychiatric Association as an area that needs 'further study'. This is the go-to manual that psychiatrists use to identify a disorder or mental health condition.

Internet addiction is being recognised globally. Rehab centres are springing up throughout the United States for young people suffering from IUD, and there are boot camps in China, where up to 30% of teens are said to have been classified as internet addicts, designed to wean screen-addicted teens off their devices. Taiwan and Korea have also long recognised IAD as a health crisis.

Internet addiction is not yet officially recognised in the UK but, when the criteria for pathological (meaning 'problem' or 'unhealthy') gambling were applied to adolescent UK gamers, researchers found one in five teenagers met the criteria for addiction.[1]

Whether or not it is an addiction, it is clearly not a good thing for a latency-age child like Ben to be showing such dependence on screen time. But he is not alone. I have seen many parents who are desperately worried that their child is addicted to their electronic device.

Why latency-age children are particularly vulnerable to becoming addicted to screen time

Screens are so addictive for children of this age because they tap into a child's need for control and order that is part of the development profile of latency. Many children of this age are obsessive about something. When they get into something, whether that is Match Attax stickers, loom bands, Pokémon cards or Shopkins, they *really get* into it. They want to collect all the cards and have the full set of stickers or toys. So if you throw a game like Minecraft or Pokémon Go into the mix then a latency-age child can get obsessed very quickly. The dangerous thing about screen time is that it is constantly available. Children don't have to wait to go to the shop after school once a week or save up their pocket money to obtain it. It is immediate, and in all our homes, which means it can easily become a problem.

Computer games and apps feed off this obsessional element. They are designed to keep people engaged – and they have no end point. They are very reward-driven: you have to complete different levels or tasks and collect things to win prizes or virtual stickers. Even educational apps have a game-play element to them and are designed to be as stimulating as possible so children get hooked very quickly.

Why screen time is so addictive

Scientists have found that playing video games causes the brain to release dopamine. Dopamine is the so-called 'pleasure chemical', and it stimulates the brain's reward system. A recent study showed that video gaming releases a comparable amount of dopamine as drugs such as amphetamines and Ritalin.[2] In fact, neuroscientists have compared the lure of screens to the lure of drugs for an addict. Researchers found that when regular video game players saw images from their favourite game, their brain responded in the way that drug addicts' brains respond when they encounter cues that remind them of their drug.[3] This addictive effect has even led to Dr Peter Whybrow, director of neuroscience at University of California at Los Angeles (UCLA), describing screens as 'electronic cocaine'.[4] My big concern here is that exposing young children to screen time that causes a constant dopamine release means that, over time, they will no longer get the same level of enjoyment or rush from playing the game, so they will have to do more and more to get the same effect. It becomes a vicious cycle.

Neuroscientists are also anxious about the effect of screen time on the brain. Researchers have found that long-term internet addiction (identified by excessive internet use) may actually rewire the way the brain works.[5] These are certainly not the effects we want for the developing brains of our young children.

Studies show that internet addiction is associated with structural and functional changes in the brain regions involving emotional processing, executive attention, decision-making and cognitive control.[6] High or excessive levels of internet use can affect the way the brain is able to make sense of emotions, focus and sustain attention and concentration, weigh up information to make decisions, and overall manage the way we think. Children's brains, which are still developing, could be very negatively affected by too much screen time in childhood.

Addiction or obsession?

I don't believe the vast majority of latency-age children are facing issues of addiction. In my experience, this issue relates much more to obsession rather than a clinical addiction. We now understand that much of what underpins addictions relates to habit, behaviour and associations. If you talk to an alcoholic they will tell you they get a feeling of release when they hear a wine bottle being opened or the sound of a can being opened, before they have even tasted the alcohol. That's an association. If people want to give up smoking, then they are more successful if they stop going to the places where they usually smoke and stop seeing the people they usually smoke with. It is hard for them to be in the environment they associate with smoking as this triggers all the physical symptoms of craving a cigarette, but this is a psychological craving, not a physical one.

There is a powerful argument that addiction is not purely chemically driven; it's actually psychological and is driven by the environment. This was shown in the Rat Park experiment, in which rats were given a large cage filled with food, toys and lots of other rats to play and mate with, while other rats were kept in solitary confinement in small cages. In the experiment, the rats could choose to drink from either a dispenser containing tap water or one containing a morphine solution. Researchers found that the rats in the small isolated cages drank much larger doses of the morphine solution – about nineteen times more than Rat Park rats in one experiment. The Rat Park rats consistently resisted the morphine water, preferring plain water. Even caged rats that were fed nothing but morphine water for fifty-seven days chose plain water when they moved to Rat Park, voluntarily going through withdrawal. None of the rats living in Rat Park had anything that resembled an addiction. Based on this study, the team concluded that contributing factors to an addiction may lie in the person's environment.

The Rat Park experiment shows us the importance of environment. Today, we are living increasingly socially isolated lives. If everyone in a family is on a digital device, then a latency-age child will copy them, and that can be very isolating. We know that people who have rich, rewarding, fulfilling lives are less likely to develop addictive or obsessive behaviours. Some children are trapped in a vicious circle when it comes to screen time. It is not the screen time that is making them feel down – but if they feel down, they turn to more screen time as an escape. The more disconnected children feel from their family and society, the more they turn to the device – until the digital world begins to take up more and more of their time. That is a fast track to an obsessive cycle of behaviour.

How to break the obsession

Whether we like it or not, screen time is part of all our lives now. We must make sure it doesn't take over our children's lives. Problems occur when screen time replaces people and real-life interactions. So if your child is overly reliant on a device, that means they are less connected to you. In latency, that is a real problem – online time isn't a replacement for real-life interaction. If you feel your child's screen time has got out of hand, then I would strongly advise parents to (1) take the device away completely, deal with the tantrums and the stress that result, and then in the future reintroduce screen time, with a schedule, or (2) reduce screen time. If that doesn't work, then take it away completely.

I don't feel it is feasible to enforce a permanent screen ban on our children, purely because a lot of schoolwork is done electronically and they are going to have to use screens throughout their lives so they need to learn how to develop a healthy relationship with them. Taking screen time away will not solve the underlying problem either. Parents need to try to understand why their child became obsessed in the first place – and what they are going to do instead.

You must decide which approach will work best for you and your child – and follow through with your decision.

I always tell parents there is no 'one size fits all' approach; it depends on your child and what the issue is. If, for example, you have a six-year-old who is obsessed with their PlayStation, and you feel they are too young to cope with it, then by all means put it in the loft for six months or until you feel they are mature enough to manage it. If, however, you have a ten-year-old who is obsessed with a tablet but who also uses the tablet for schoolwork, then it is not practical to take the screen away permanently. You are going to need to work on finding ways to manage it instead.

Going cold turkey

When a child's obsession with screen time is severe and parents have lost all control, then sometimes the best option is to take away all digital devices. When I suggest this to parents, they always look horrified. They are genuinely terrified about what might happen and how their child will react, and they will often make excuses for why they can't do this.

Banning screen time will require strength and resolve. Your child will probably have tantrums and arguments – but stick it out and it will pass. Don't be afraid to turn off the Wi-Fi and remove the power supply/chargers if you need to. The important thing is to be firm and consistent and not give in to moaning or pleading. It generally takes five days to break a habit or a pattern of behaviour, so the backlash will not last forever.

How long you ban screen time for is again an individual decision based on your child and the severity of their problem. I think it needs to be for at least one to two weeks to allow your child to adapt to a screen-free life and begin to engage in new activities and interests, or take up old hobbies, to replace screen time. After that, reintroduce screen time slowly, with strict rules and boundaries in place so it does not become an obsession again. Parents must get on top of it, keep on top of it, and stay vigilant so it doesn't start to become a problem once more.

How to cope with the backlash

Many parents are worried about their child's reaction if they reduce screen time or take their child's devices away. I always use the analogy of a storm. If there is a typhoon blowing through your house, you don't go and stand in the middle of it and try to manage it. You batten down the hatches, wait for it to pass, and clear up afterwards. Then you think about how you could protect yourself from the same thing happening again next time. In the same way, if your child is kicking and screaming because they can't have screen time, then learn to ride the storm; don't feed it. Walk away – don't engage with them or try to negotiate because it will only ramp things up. If you have given them an explanation in the beginning for why you are taking away their screens or cutting screen time, there's no point in going over it again and again. They almost certainly understand why you are doing this. Their protest is not because they don't understand, it's because they don't like it – and no amount of explaining is going to solve that.

Many parents worry about the consequences of their child's meltdowns or tantrums. They worry about their child harming themselves and things around them. However, it is important to keep this in perspective. There is only so much damage a latency-age child can do. If they are screaming and shouting and slamming doors, the worst they are (generally) going to do is give themselves a headache – and the minute you pay attention to them it is giving them a reaction. So it is vital that you ignore them unless you think they are in danger – then you step in swiftly and deal with it.

A lot of parents are fearful of doing this, and find it very difficult. It is like going back to when your child was a toddler having a tantrum. Often the best way to deal with tantrums is to ignore them and walk away. The same applies to a latency-age child. Your child is angry because they can't have something they want, and nothing you say or do is going to appease them. They need to know you will stand firm. Distract them, move them on, try to divert their attention elsewhere. When they show more positive behaviour, make sure you give them lots of positive attention. This way, they will quickly learn that their positive behaviour will gain your positive attention, and they will prefer this to negative attention.

Case study

I wanted to talk about this particular case because it is an extreme example of what can happen if parents lose control in latency. It shows how a child's obsession with the online world can grow to harmful levels by the time they reach adolescence. By the time fourteen-year-old Joe was referred to me, he was on his Xbox for 19 hours a day and his parents had lost all control. The Xbox was in his bedroom and he was gaming throughout the night, falling asleep for a few hours in the morning and not getting up in time for school. His personal care was non-existent – he wasn't showering or eating properly, and at the peak of the problem he was actually defecating in takeaway containers in his bedroom because he did not want to leave his screen, even for a few minutes. The situation had become totally unmanageable, and social services had become involved because of Joe's lack of school attendance, lack of self-care and genuine concern for his safety and well-being. His parents were desperately worried, and felt that the situation was out of their control. Social services had made threats to take Joe into care if he did not start going to school.

THE SESSIONS

Understandably, Joe's parents were desperately worried. I talked to them about what had led up to this point. Joe had always been into gaming and digital devices. It was clear his parents didn't really understand the appeal, but were not particularly worried about him generally, so he had largely been left to his own devices to be on them when he wanted. Joe had been very keen to have an Xbox and he had been able to buy one at the age of ten with Christmas and birthday money that he had saved up. The Xbox had then been installed in his bedroom. He was quiet and not causing any trouble, so they didn't see any problem with his screen time. His time online had gradually increased over the years, but the crunch point came when he struggled to cope with the transition to secondary school. He hadn't got a good group of friends and he found it academically challenging. From that point onwards, Joe spent most of his time in his room on his Xbox. His mum noticed that he was tired all the time as he had been up late gaming. He was good at gaming and had a group of online friends – a lot of whom were in America – so the time difference meant he stayed up through the night, playing and chatting to his new friends. His mum had tried to put limits on this, but Joe had made such a fuss that eventually she had given in.

The harder things got at school, the more Joe had retreated into the online world. It was a quick downward spiral. By the time he hit puberty, he was up all night gaming, and his day/night patterns had reversed. He would fall asleep in his gaming chair and then not be able to get up for school. He was 5ft 10in and his parents could not physically make him go to school. An educational welfare officer had become involved, and the local authority was threatening legal action against his parents.

When I went to see Joe at home, it was clear that his mood was incredibly low. He wasn't eating properly; he wasn't getting enough good-quality sleep or natural light. He was spending all day in his room with the curtains closed, on his Xbox. He wasn't going to school and was socially isolated. He had retreated into the online world as he felt totally disconnected from – and unable to deal with – real life.

INTERVENTION

Drastic action was needed immediately so his parents could regain control. Joe was stuck in a vicious circle. The first thing his parents had to do was take away the Xbox. Even though things had become desperate, they were still unsure about this and they were worried about Joe's reaction. In this extreme situation cold turkey was the only way to go because Joe had gone past the point of timetables and restrictions.

This time his parents presented a united front and removed the Xbox from the house. As expected, Joe reacted. There was a lot of door slamming, shouting and storming off.

However, once he knew that everybody was behind this and that his parents were backed up by an army of professionals who were all saying the same thing, he gradually accepted it. I think part of him was actually relieved because he was not happy with his life. His mood was very low and he was miserable.

One of the first tasks was for Joe was to rebuild his links with school, and his relationship with his parents. Once the Xbox had gone, we had to try to persuade him it was worth getting up and going to school. His parents worked closely with the school, and Joe was given a mentor and was taught on a part-time basis initially in an additional needs unit.

OUTCOME

It was a long, hard process for Joe, but after several months the situation had improved and social services was no longer involved. He was still struggling academically, but was being given extra support and he had made a couple of friends. Without the gaming, he wasn't staying up late so he was sleeping more and his parents were able to get him up in the mornings. After four months, at the end of the academic year, his parents reintroduced screen time – but this time with restrictions. He was only allowed to play on the Xbox downstairs and at weekends. Joe stuck to the rules, as he knew that if he didn't the device would be removed again. He also had less interest in gaming as he wanted to meet up with his real-life friends. Thankfully, he had begun to realise there was more to life than sitting in his bedroom in the dark.

Red flags

❶ Would you describe your child as being 'obsessed' by their device? Are they constantly nagging to be on it? Do they think about it all the time?

❶ Do they become upset and anxious at being separated from their device?

❶ Do they ignore instructions to come off the device and become very distressed when forced to do so?

❶ Is the device the sole focus of all of their conversation, games and behaviour? Even when they are not on the device, are they talking about it and what they are going to do next on it? If you're out, do they seem anxious and want to go home so they can be on the device?

🅠 Does your child seem stressed? (When you are obsessed by something it is very stressful. When your child is having screen time, they are working hard to feed the frenzy and when they are off it, it is all they are thinking about, which is exhausting.)

🅠 Is your child defensive or secretive about their screen time? Do they hide their device so you cannot find it, sneak it into their bedroom at night, or lie about the amount of screen time they are having?

🅠 Are they constantly tired, irritable and withdrawn?

Some solutions

➲ Go cold turkey. If parents are seriously concerned and want a quick fix then take devices away and ban screen time.

➲ Either when the child is older or parents feel they and the child can manage their online time better, reintroduce screen time in very small doses.

➲ All parents should have a plan in place for their child's screen time. Come up with a timetable – either set times the child can be online and schedule it into their day, or allocate them a certain chunk of time they are allowed to be online for. Once they have used that up, there is no more for that day or week.

➲ Parents have got to police the timetable if they want to avoid their children becoming obsessed.

Chapter 6

Social media and age restrictions

**Selfies, social media, and how to teach
your child to behave responsibly online**

A friend confided in me recently that her ten-year-old son Matthew had got into trouble because he had posted a photograph of himself on Instagram with a caption saying: *the first girl to like this will get a kiss from me and can be my girlfriend.* 'Some of the girls in his class had seen it and one of their mothers rang me to complain,' my friend told me. 'I was mortified, partly because I didn't even know he had an Instagram account on his mobile phone. I didn't have a clue what he had been posting, who he had been following, or who was following him. He didn't even have a protected account – so anyone could see his photographs.' My friend was also unaware that social networking sites come with age restrictions.

The minimum age for Twitter, Facebook, Instagram, Pinterest and Snapchat is thirteen, and for YouTube it's eighteen. However, a thirteen-year-old can open an account with their parent's permission. The reason for a 13 limit on most social media is due to the US law Coppa (Children's Online Privacy Protection Act), which requires online services to seek parental consent from users below that age. The issue of consent is also to do with protecting children: the restrictions are there because these social media apps are not suitable for a young child like Matthew.

My friend isn't alone. A recent study found that 53% of parents are unaware that social media sites like Facebook require users to be over the age of thirteen. Worryingly, one in five parents thought there were no age requirements at all.[1]

Despite this, statistics tell us that more than half of children have used an online social network by the age of ten.[2] A recent survey of 1,500 parents by Internet Matters found that six-year-olds are now as digitally advanced as ten-year-olds were three years ago. Around 44% go on the internet in their bedrooms, and 41% do so unsupervised. The survey also found a third of six-year-olds also use WhatsApp, and a quarter are now on social media.

It's important to understand that these age restrictions are there for a reason. The Office of National Statistics' report on children's well-being found that children who spend more than three hours each school day on social media sites are more than twice as likely to suffer poor mental health.[3] The report stated that this amount of screen time may cause these children to experience delays in emotional and social development. Researchers have found that other negative effects of social media are increased exposure to harm, social isolation, depression and cyberbullying.[4] A report published by IZA Institute of Labor Economics even suggests that just one hour a day on social media can make a child miserable. What is so alarming about this research is that it is telling us about the impacts of social media on adolescent children – but we know that younger children are even less psychologically equipped to deal with these challenges.

It is not unusual for me to see six-, seven- and eight-year-olds who have their own Facebook and Instagram accounts, and the majority of children at the upper end of latency (ten- and eleven-year-olds) are using some kind of social media. One of the most worrying things about this is that through social media we are bringing adolescent issues to latency-age children – and expecting them to cope with this without the developmental maturity that comes with adolescence. This is very serious. A young child is just not developmentally able to manage the complex themes that begin in adolescence, such as intimacy and sexuality. Even when parents think they are managing this issue – for example, they know their child's passwords or they only let their child log in via their own phone – latency-age children can be sneaky. They can soon work passwords out or log in using a friend's device.

During latency, a child prepares mentally and physically for the challenges of adolescence, which include the development of sexual awareness, body image, establishing identity, and becoming an individual. Primary-age children don't have the cognitive or emotional resilience to handle social media sites – particularly those that focus on appearance and social competitiveness – such as Snapchat, Facebook and Instagram. Being on social media too early could expose them to people, content and situations they are not prepared for and they might feel out of their depth. Neuroscience tells us that adolescence is a crucial time for the development of the brain's prefrontal cortex, which is the part that helps us form judgements, control impulses and emotions, and functions like decision-making, self-awareness and understanding others. Before this develops, latency-age children are much less likely to have the emotional regulation and resilience necessary to cope with the things they might see or experience in online environments that are not designed for them.

Exposure to social media can even affect education. One survey found that, according to teachers, children with the poorest grades at school are the ones who spent most time on social networking. Half of the 500 teachers polled believe their fixation affected children's ability to concentrate in class, and two-thirds said the quality of children's homework was poor as they rushed to finish it so they could communicate with friends online.[5]

As adults, we are fully aware that social media can be addictive and my main concern is that children are particularly vulnerable to this addictive effect because of the tendency towards obsessiveness during latency. Researchers at UCLA's Brain Mapping Center found that being appreciated in social media through 'likes' was seen (in brain scans) to activate the reward centres of the brain. Children's brains responded in a similar way to seeing loved ones or winning money. In my clinics I have seen nine- and ten-year-olds whose parents have told me they are obsessively checking their Facebook or Instagram feeds to check the number of 'likes' their posts and photographs have got, which seems to help determine their social status in the playground. A low number of 'likes' typically translates into low social status, and possible shaming and bullying. A high number of 'likes' might mean a child is popular – but then children feel the pressure to sustain that status.

I have also spoken to worried parents whose children get stressed about breaking their Snapchat 'streak' – the number of days they have messaged a friend unbroken. Some children have 'Snapstreaks' with over 50 people, which means they are on the site for hours every day. A friend recently described to me how her daughter had become hysterical on holiday when she had realised there wasn't reliable Wi-Fi in their apartment. When she finally managed to access Wi-Fi in a nearby town, my friend was shocked to discover her daughter had received over a thousand Snapchat messages in 24 hours. It made her seriously concerned about how much time her daughter must be spending on social media if she was used to receiving and reading so many messages every day. Latency-age children should not be facing these kinds of pressures and challenges.

The American Academy of Pediatrics has even warned parents about something they term 'Facebook depression', which is experienced by children and teens when they see a status update, a wall post or a photo that makes them feel unpopular.

Latency-age children and mobile phones

Children are doing things earlier and earlier. Today, some previously adolescent situations such as having a mobile phone are trickling down into latency. As recently as five years ago, it was common for children to get a phone at eleven when they started secondary school, but I am now seeing four- and five-year-olds with their own mobiles. One study by the National Literacy Trust found that 79% of children aged seven to eleven had their own mobile phone, and that children were more likely to own a mobile than a book.[6] I have seen this shift in mobile phone acquisition unfolding in front of me, and I remain unconvinced about the need for a young child to have a mobile phone. The majority of phones today are smartphones which means as soon as a child is given one, they are being introduced to social media.

Body image and selfies

Young people today are selfie-obsessed and latency-age children, girls in particular, using social media will often post photos of themselves online and invite people to comment on their appearance. I cannot stress enough how unhelpful this is for young children. I regularly have to deal with distressed young children because they have posted an image of themselves that has received a negative reaction, or they've seen an image of themselves they have not been happy with.

If a latency-age child is using social media and they want to post selfies, then I think it is important for parents to talk to them about why they feel the need to express themselves in this way. Is it about keeping up with other people? Is it about fitting in and following the crowd? Or is it about exploring or playing with their identity?

If a child is going to use social media, it is important for parents to help make them critical consumers, and for them to be aware that in the online world things are not always as they seem. It is incredibly refreshing when I hear a group of teenage girls looking at photos on social media and talking about how Photoshopped and filtered they are. Adolescents have the emotional maturity to see through filters and know that someone does not actually have flowers in their hair and a rainbow coming out of their mouth in real life. Latency-age children are very literal and honest. They will believe what they see, and they haven't got the cynical, critical eye of a teenager. You need to give your child the skills and resilience to cope. Look at Instagram or Facebook with them and talk about what you see. Talk about the way people can easily manipulate their image, and discuss things like filters. Children are more aware of image than they have ever been before, thanks largely to smartphones. Having the internet as well as a camera in your hand all the time is a powerful combination, and it only takes seconds to upload a photo.

It is important that children are made aware from a young age that images used in adverts and in magazines are Photoshopped and airbrushed: a model's skin isn't really flawless and an actress's legs or waist are not really so thin in real life. In the past, actress Kate Winslet spoke out after a cover she did for GQ magazine was excessively airbrushed. She has also stated in her contracts with cosmetic giant L'Oréal that her photos used in their campaigns cannot be digitally retouched, because she feels she has a responsibility to a generation of young women. What worries me most is that we really don't know what effect this level of image perfectionism is having on children who are at a vulnerable and impressionable stage of development. Recently, psychologists found evidence linking social media use to body image concerns, dieting and a desire for thinness in adolescents. Social media platforms such as Facebook, Instagram and Snapchat allow teenagers, and increasingly children, the opportunity to compare themselves to others and earn approval for their appearance. I know several ten- and eleven-year-old girls who regularly post selfies online, inviting people to comment – and therefore opening themselves up to criticism.

I recently treated a twelve-year-old girl who was hospitalised after developing an eating disorder. Her body image issues had started at the age of eleven when she had got a smartphone and joined in group chats on WhatsApp and Snapchat with her friends. As many young girls do, they continually posted pouty selfies and this girl had received some very critical comments on some of her pictures from girls in her class who said she was fat and her thighs were chunky. They had then started sharing the photo with other members of the year group. This girl got very upset and started to respond to the comments – and because they knew they were getting to her, they carried on. In response, the girl started to post more and more selfies, desperately trying to prove she wasn't fat. She just couldn't stop. The bullies sensed her weakness and vulnerability there and kept on making bitchy comments. It had badly affected her self-image and perception of herself. She was making herself vulnerable and inviting criticism from others. Her image of herself became completely distorted. She saw herself as fat and developed an eating disorder in order to try to address that. It took her family a couple of months to realise what was happening. By then she had lost a lot of weight and she was admitted to hospital. It was all triggered by what she had experienced via social media.

At primary school age, it's not about teaching children to cope with this continual comparison; it's about restricting it for as long as you can until they are ready. Why do young children feel they need to post images of themselves all over the net? It's essential for parents to prevent this from happening. It can provoke obsessive behaviour. Someone posts a picture and you like it and they like yours – and it encourages children to start posting more and more. Parents of a latency-age child who wants to use social media in this way should limit the number of photos their child is posting a day. Look through social media with them and take an interest in who they are following and why. Teach them to be a critical user. Tell them that glamorous bloggers are paid thousands of pounds to promote products and to look the way they do, and that celebrities don't look the way they do naturally; they have stylists, make-up artists, personal trainers and plastic surgeons.

Encouraging digital responsibility

It is crucial that, from a young age, children take responsibility for their online behaviour. They need to understand that the internet is a permanent record and their digital footprint lasts forever. I always tell my children not to post anything online they would not want me, their teacher, or a future employer to see. They need to realise the internet is a public arena, not a private one. You can find out all kinds of things about people by doing an online search of their name, and you need to take control of what people can find out about you. What goes online, stays online.

As parents in the digital age, we have a responsibility to teach our children how to manage their online identity just as they do their real-life identity. Most parents are very clear about what they would and would not let their children wear or walk around the streets in, but do we apply the same rules and structure to their online identity – when in fact their online image is going to reach many more people than they will encounter in the real world?

Perhaps even more important is the fact that their online image will last forever. Your online history can even affect your employment and career prospects, as many employers and university admissions officers will look at a candidate's social media profiles. One survey found that 60% of employers look at a candidate's use of social networking sites when considering who to interview. It also revealed that more than 30% of the employers surveyed use social networks to research current employees. Of those employers, more than 25% have found content that resulted in them reprimanding or firing an employee.

How to talk to children about the news

One mother told me how her seven-year-old son had been using the family tablet when a BBC news alert popped up on screen, describing how a seven-year-old girl had been stabbed. Her son was incredibly distressed, and it was hard for her to reassure him. A lot of the news we hear today *is* very distressing. We can't understand it ourselves, never mind reassure our children. So what can parents do when the news is so accessible online? All it takes is one click on a tablet or phone to get them to the headlines, or children might read things on social media about news and current affairs. When there is a terrorist attack, for example, there is no way of avoiding it on social media. So should we actively try to keep a latency-age child away from the news, or is it unrealistic to try to shield them from the world we live in?

Children of this age may become fearful and confused if they are struggling to make sense of the news or other adult topics without guidance. That's another reason why it is so important that, wherever possible, parents should not leave a latency-age child to browse the web alone. If they do come across something that upsets them, you could reassure your child by explaining that the fact an incident has made the news means it is rare. It is important to talk to your child about what they have seen and what they know about an event to check they have got their facts correct, as young children can often misinterpret what they have read. Ask them how it made them feel when they read this story, what they think about it, and what they are worried about. It is good to be honest with our children and admit that some of the news is very scary and that we can't control everything that happens. Children aged about five to eight will want to know more about how what has happened affects them, whereas older children (from nine to eleven) may be able to see the wider issues. All latency-age children will want to be reassured about their safety.

Following a terrorist attack, for example, you might explain all the things that happen afterwards, such as extra police on the streets and increased security measures at public places like train stations, stadiums and airports. You might also want to point out all the people who ran to help, such as passers-by and the emergency services. Say there are more people in the world who want to help than there are people who want to harm. Give your child the opportunity to ask you questions. What has the news made them think about? What does it mean for them, and your lives? What could you do to help? Take the conversation away from the actual event – which none of us have any control over – back to your life and your family.

Working out the rules

Talk to your children about social media and work out some rules you think are right for them and your family. Neither of my daughters had social media accounts before they were thirteen, and I was quite firm about that. I explained to them about the age limits, and they were given the same message by their school, which helped. I was also very strict about the kind of images they were allowed to post on social media when they were old enough to have their own accounts. They were not allowed to post pouty selfies or photographs of themselves online. As a family, we talked a lot about how they would like the 'world' to see them – and that meant both the digital world and real world. We spoke about ideas of identity, and that we can all have different identities in different settings. For example, I am different at work and at home. My children have all visited me at work at one time or another, and think it's very funny to see me in work mode relating to my colleagues and managing my workload because to them I am Mum, who charges round the house in pyjamas with wet hair, shouting that we're going to be late. This is a very clear example of the different identities we all have; not everyone sees all the different versions of us. I explained to my children that on the internet it's very easy for those boundaries to collapse and for people to show their personal or private identities to the whole world – and I do mean the whole world. It's essential for children to realise that the internet is global. It's everywhere – and it's forever. Once something has been posted on the internet, it's impossible to get it back. My colleagues will never see the glory of my bedraggled, pre-work, drill sergeant, morning mum routine – and there is a very good reason for that. But thanks to smartphones with video cameras and YouTube, the whole world could witness it at the click of a button.

Even as my daughters have grown up, they have still chosen not to show their faces on social media. My daughter is sixteen now and her Facebook profile is a photograph of her silhouette in the sun. They have grown up with me asking them if they want people to have access to their image. They know they can Google my name and find lots of images of me, because of my job, and that has made them feel quite fierce about their own privacy. I think it's very important to ask yourself how you portray your own identity online, and to think about what messages that is sending to your children. Based on your own views about privacy and the portrayal of your identity online, how do you want your children to show themselves to the world?

Don't just restrict and ban

From a professional perspective, I don't believe a latency-age child needs to be on social media sites. However, simply restricting and/ or banning them is not the answer. You need to have a conversation with them about social media when they're young, because they will be on it one day and you need to make sure you give your child the skills to cope with it. Talk to them about what age you think it is appropriate for them to have social media accounts, and agree that together. It is far easier to negotiate beforehand than it is to get them off social media when they are already using it. Social media is an important tool for teenagers – and it is part of their social life. By all means restrict it if they are too young, but educate them, talk to them about it, and help prepare them and build their resilience. Help them spot the difference between what is real and what isn't – and be discerning about what they look at and what they show. Sit with your child, talk about social media with them, and look at the things they are looking at. Use restrictions, but make sure you talk to your children about how to navigate the online world. Talk openly and straightforwardly about the risks they may encounter online – without scaring them.

The recent 'Growing Up Digital' report by the Children's Commissioner[7] called for earlier and wider technology education in schools, and recommended that this should be expanded to include the 'social' elements of life online. I firmly agree. Online safety, digital responsibility and coping with social media should be as much part of the curriculum now as maths and literacy.

Loneliness

Somewhat ironically, social media appears to be exacerbating feelings of loneliness and isolation in children.

One study by Childline found that social media leaves children feeling isolated, with many saying it led to them comparing themselves to others and feeling inferior, ugly, and unpopular as a result. This is reflected in my own clinical work, in which I see nine- and ten-year-old children telling me how alone they feel when they look at social media. I've experienced it with my own children too. Both my daughters have had times where they haven't been included in a group meet-up of their friends or haven't been invited to a party. Before the internet, you wouldn't have known you'd been left out, but now there's no chance of avoiding it, as everything is posted all over social media. Imagine your phone pinging all day with messages and photos of your friends enjoying an event you haven't been invited to.

Each of the people at that gathering are posting images. The only way to stop it is to leave the group – but the app will state you've left the group, which looks like you're having a strop. Then you can't get back in unless someone invites you back, and if you're out of the group chat, you're out of the group, and you miss out on everything else going on in the group. Imagine how that feels. Most adults would find that intolerable, so imagine how hard it must be for a child to cope with. In this sort of situation, I tell children to turn their phone off or to put it away, and help them focus on finding something enjoyable to do that will distract them.

It's tricky for parents to know what to do in this situation, but I think generally they need to keep promoting real-life, face-to-face connections. Encourage your child to meet up with friends and have them round. Try to take things offline and back into the real world as much as possible. Real-life connections are hugely important. With young children, parents should have tough limits about them putting their phone down and having time away from the virtual world, because otherwise it becomes all-consuming and they never get any respite from what is going on. I know a fourteen-year-old girl who hands her phone over to her mum at 8pm every night to stop her being on social media all evening. Her mum charges her phone and gives it back in the morning. She moaned about it in the beginning, but she's completely used to it now.

At latency age, it's all about limits. Limit the time your child spends on social media. Take their phone from them, limit the number of group chats they're on, and explain that they don't have to respond to every message. It doesn't matter if they are not constantly online. Your phone could break and you might be off the internet for three days, but nothing bad is going to happen; there is nothing to be anxious about. You will go to school and people will still speak to you. The earlier that children start using social media, the more intense it's going to get: it's better to delay it as much as possible for this age group.

Case study

Eleven-year-old Dan was referred to me as his parents were worried about him: his mood was low, he wasn't eating or sleeping properly, and he was showing many of the symptoms of depression.

THE SESSIONS

As I tried to understand what had caused his low mood, his parents explained that Dan had had a few problems because of social media. A few months ago they had allowed him to have an Instagram account, as many of his peers had one. His parents had monitored it at first, but had been reassured as he mainly posted photos of the family cat, cakes he had baked, and his toys. Other pupils in his class had started teasing him about his posts, and mocking them for being babyish and naive. Dan was upset by this and really wanted to fit in so he decided to make himself look more grown up. In order to give himself street cred and to look 'cool' to his peers, he'd started to create an alternative persona online. He created a false account for an older girl and pretended to be in a relationship with her. Across the accounts, it looked like they had shared pictures and there were conversations between them. He'd also started approaching other girls at school to get photos of him and them together, so on social media it looked as though he had a whole posse of girlfriends. One of these girls' parents had become concerned about Dan's stalker-like behaviour: they had informed the school and Dan's parents were told. His parents were alarmed at the adult way he'd been trying to portray himself online. Everything came crashing down around him and he had to admit what he had done. Everyone at school found out and he was ostracised. Dan had been left traumatised, distressed and miserable.

INTERVENTION

I had to support Dan, but I also had to go back to the beginning and talk through how this started. We talked about how Dan wanted to be like everyone else; he said he was upset at being picked on. He was doing quite well in the real world before this had happened, and he had friends. But when the social arena had started to move into the online world he couldn't keep up, and it had spiralled massively out of control.

Initially, Dan couldn't see a way out of the situation. He felt that he desperately needed to rebuild his online life, and this was the only route back to getting 'in' with the group. In fact, the opposite was true. I helped Dan to understand that the online world was transient, temporary and not at all necessary for good day-to-day functioning. I then worked with him and his parents to rebuild his real life – the one that needed attention.

OUTCOME

Dan gradually began to re-establish social links with his class-mates, particularly those who were not online and who were interested in similar things to him. We completed some work on online awareness, and Dan helped lead a school assembly about online behaviour.

Red flags

The red flags below could indicate that your child has an obsession with, or an over-reliance on, social media. If you think your child is becoming obsessed with screen time, you must address this. You must reduce your child's screen time – or remove screens altogether if necessary. If the situation is extreme, then contact your GP, with a view to accessing additional help.

- They are continually checking their phone. They're obsessed by their smartphone, which is literally glued to their hand.

- They're constantly taking photos of every aspect of their life, from their breakfast to their pet cat.

- They're continually posting selfies online and inviting comments about their appearance.

- They're involved in large numbers of group chats or Snapstreaks and their phone is always pinging.

- They get very anxious or distressed if they are away from their phone or if they can't access Wi-Fi.

Some solutions

- Social media is essentially an adult environment: it is not a suitable space for latency-age children. Most platforms have age restrictions, so follow these guidelines wherever possible.

- If you do want to give your child access to some social media then you need to police it. Make it clear that you will have access to their Instagram or Facebook account and you will monitor it.

- If you feel that social media is taking up too much of their time then limit the amount of group chats or 'streaks' they are on, or the number of photos they're allowed to post.

> Talk to your child. Come up with rules you agree on – this will make them feel more involved in the decision-making. Discuss what apps they want access to, and look at them together. If you disagree with them, then explain why. Ask them what they think is OK for children of different ages to look at.

> If your child wants to look at news and current affairs, then find an age-appropriate news site, such as BBC Newsround (www. bbc.co.uk/newsround). Look at the site with them and talk about what they have read. (It is a good idea for parents to read articles themselves first to check they do not contain any graphic or upsetting content.)

> Lead by example when it comes to social media. If you are always taking selfies and posting them on Facebook and Instagram, then your children are going to want to do the same. Your behaviour is going to filter down to them, and the first thing they will want to do when they get a phone is to start posting selfies because that is normal to them.

> Be aware that once your child has a social media account they will have access – and a way of signing in – to other websites. You can sign in to many websites via Facebook or Twitter.

Chapter 7

Online risks and keeping children safe

Grooming and cyberbullying – how to minimise the risks and keep your child safe online

This is the number-one worry for most parents I speak to. If a latency-age child is online, how do we keep them safe? The statistics are deeply disturbing, and suggest that parents are taking their eye off the ball. One Childwise survey found that three out of four children aged five to sixteen (73%) had internet access in their bedroom, and 10% of them had no restrictions on who could see their personal details online.[1] A report by the NSPCC found that almost a quarter of eleven- and twelve-year-olds who had a profile on a social networking site had been upset by something they had seen on it in the past year. Worryingly, over half (62%) of these experiences were caused by strangers, someone they only knew online, or the children did not know who had caused it. When they had been upset or bothered by something on a social networking site, younger children tended to be less assertive than older children – another sign that a latency-age child does not have the social maturity or resilience to cope with social media.

The worrying thing for me here is that parents may not be keeping as close an eye on this as they need to. Just 32% of parents described themselves as feeling 'very confident' about helping their children stay safe online.[2] We supervise our children far more in the real world than we do online, yet the internet is one of the most genuinely dangerous places for them. In the real world we tell children not to talk to or go off with strangers – however, often we don't replicate that online, because we can't physically see the potential risks.

I'm still constantly amazed by the naivety of parents who allow their young children unrestricted, unsupervised access to search engines and websites. I recently spoke to one mother who had allowed her seven-year-old son to publicly post a video on YouTube of him and his sister play-fighting. She had looked at the video before he had posted it – but she was shocked to see a stranger had posted a series of sexually explicit comments on the video.

Devices are a one-to-one experience, unlike watching TV or watching our children playing. When a child is in front of a screen, parents generally don't know what they're doing. They leave them to get on with it because they are quiet – which is unlike any other situation in the real world.

Vicki Shotbolt, the CEO and founder of Parent Zone, describes the online world as being 'like another high street for kids', stressing that we need to be in it and parenting in it. I couldn't agree more. Parents need to start realising that, when it comes to screen time, we must be vigilant.

Virtual friends are not real friends

A latency-age child should not be chatting online to strangers, but sometimes, despite a parent's best efforts, they may come across people they don't know in a social forum or group chat in an online game. It is important to make children realise that if they don't know someone in real life then they are still a stranger, even if they have talked to them online. This is very hard for young children to understand.

Digital relationships can become intense very quickly. I come across a lot of children, some as young as twelve or thirteen, who say they've had romantic relationships with people they've met online. Most of the time they're not in the same country and have never met or even spoken. The whole relationship is conducted via messaging or emails. Children say things to digital friends that they wouldn't dream of saying in real life, and disclose personal information very quickly. Parents must explain to their children that digital conversations are not the same as real-life conversation, and you don't really know who you are chatting to behind a computer screen. Before they respond, children need to ask themselves, would they be comfortable having this sort of conversation face to face with someone? Would they share these personal details with someone they had just met in real life?

Online grooming

Grooming is a process during which someone prepares a child for sexual abuse. Unfortunately, the anonymity of the online world provides many opportunities for grooming. A person might pretend to be a child and use a child's profile picture. They might befriend a child on social media or in a game. They might engage the child in online chat, and share stories they know they will be interested in or chat about hobbies or common interests. It's all about building a relationship with the child and getting the child to trust them. It's important that any child using the internet knows about grooming, but parents are often worried and confused about how to talk to young children about the dangers.

The most important thing a parent can do is transfer the idea of 'stranger danger' into the online world and talk to children about grooming (in an age-appropriate way). Educate them early and often. Explain that, in the online world, people are not always who they say they are. Children often view social media as a popularity contest and think the more friends or followers they have, the better. But a latency-age child should never accept a friend request from someone they don't know in the real world. They, preferably along with your help, should also double-check requests they get from someone they think they know, to make sure it is genuinely that person's account. Is that person who you think they are?

The same applies to messaging and chatting. As a rule, a latency-age child shouldn't be chatting with strangers online and should never agree to a private chat with a stranger. Teach your child that if a stranger sends them a message or tries to chat virtually with them then they *must* let you know – just as they would if someone approached them in the street or in a park. Even if your child doesn't browse the web unaccompanied by an adult (as I recommend), you still need to have these conversations with them. Children must be web-wise. Be wary of multi-player settings on games where children can play with strangers. As I mentioned above, latency-age children should only be playing and interacting online with people they know in real life.

Talk to them about the information they give to people when they are online. A latency-age child should not share their full name, their home or email address, their phone number or the name of their school with people they do not know in the real world. Make sure they pick a username that doesn't give away any personal information and is no way similar to their real name. Explain to them how everything they share online – like usernames, images and comments – builds up a picture of who they are, so people might pretend to know them. If a stranger came over to your child in the playground and asked them for their name or address, there is no way they would give them that information – yet they may do in the virtual world.

The most important thing is for parents to talk to their child and be vigilant. Get them to show you the sites they are using and the things they're doing online. Explain to them that when they are online they need to act like a detective all the time. How do they know that person is really the age they say they are? What do they know about them? Have they ever seen a picture of them? Are they sure that's their picture? Teach them that the virtual world is not the same as the real world, and they can't just accept things at face value. In the real world we know who teachers are and who police officers are. We can see where they work and the uniform they're wearing.

The other week, my son wandered off in the school playground and lost sight of his childminder, who had gone to another playground. He got lost, but he knew to approach one of the other mums, who helped him find the childminder again. I asked him how he knew who to pick, and he said he'd seen me talk to this mum and he knew that she knew me so she'd know who he was. Children use these discerning powers all the time. We teach children to look all the time – look for somebody you recognise, look for an adult who's wearing a name badge or a uniform. But there is nothing to see online, and we don't know who is a safe person if we get lost. In the real world, children might be nervous to approach an adult, whereas online they will merrily chat to anyone in a game. We can't use our judgement skills online because nothing can be checked or verified. So we have to teach children that there's no way of knowing who people are. The safest message is: don't chat with someone you don't know. You need to encourage them to come to you if they're stuck or unsure about something.

Sharing inappropriate images

I know that children as young as ten or eleven are sharing inappropriate images online, but parents are burying their heads in the sand. They assume that it's a teenage problem, that their children wouldn't do this and it won't happen to them – yet it is. It happens frequently. Children of this age do not have the emotional maturity to cope with the consequences of circulating or sending explicit images of themselves, and I see children whose lives have been shattered as a result. The other day, I was talking to a fifteen-year-old boy about inappropriate images. He told me that every single boy in his year had sent or received a sexually explicit picture. That shocked me – but for him it was a normal, everyday occurrence.

If you're giving your child a phone with a camera, then you must lay down the rules about using it before they get their hands on it. It sounds dramatic, but you are giving them the means to make some devastating and irreversible actions. Most children inherently know it's a bad idea to share explicit images of themselves, but they get into a chat with another child and they mistake this sort of request as intimacy or some sort of dare or game. They act without thinking about the consequences. The sad fact is, once you have sent an image of yourself to someone you have lost it. The recipient has an image of you – and they can do what they like with it. Girls will often share breast or cleavage shots. Boys see these as some kind of trophy, and want to show their girlfriend off to their mates. Before they know it, the girl's picture is all around the school. That can be devastating.

It's only when children get caught out that they realise how permanent the internet is. Even a Snapchat can be screenshotted before it disappears. I've been working with a teenage boy who, when he was twelve, was persuaded by a group of girls to send them a photograph of his penis. They shared the photo with others, who circulated it around the whole school. This boy was bullied and mocked – and he ended up not attending school for two years because he was so humiliated. He couldn't get over what the girls had done to him. He kept going over and over it in his mind and he couldn't let it go. That one poor decision affected his entire adolescence.

I cannot stress to parents enough that this is real, it is happening, and it needs to be addressed. Children need to think about the images they share virtually, and they need to know that once something is out there it's very hard to remove. If parents suspect their primary-age children are sharing or receiving inappropriate images, then they need to act straight away. They must explain to their children that there's absolutely no need for it. It won't make people like them more; instead, it will make them vulnerable, and their photo could be shared with lots of other people. If children are in this position then they need to speak to a grown-up and ask for help. For latency-age children, there is no room for negotiation here. It's a definite no.

Cyberbullying

Childline has seen an 88% increase in calls about online bullying over the past five years. Some of these calls have been about children as young as seven. Nearly a third of these said the abuse had happened on gaming or social networking sites.[3] One US study found that 87% of the 1,502 children aged between ten and eighteen they questioned had witnessed cyberbullying.[4] The impact of cyberbullying on young children can be very serious. Latency-age children do not have the emotional resources to cope with the devastating effects of being bullied, especially the kind of critical or humiliating bullying that occurs via social media. Often online bullying involves criticising a child's posts or pictures. This type of bullying is heavily shame-based, and can have tragic psychological consequences. When children are being cyberbullied they feel there is no escape. The bullying is not confined to the playground; it can happen wherever they are, day or night. There is no safe space.

Cyberbullying includes:

✪ sending threatening or abusive text messages

✪ creating and sharing embarrassing images or videos

✪ 'trolling' – the sending of menacing or upsetting messages on social networks, chatrooms or online games

✪ excluding children from online games, activities or friendship groups

✪ setting up hate sites or groups about a particular child

✪ encouraging young people to self-harm

- voting for or against someone in an offensive poll (such as who is the ugliest girl in your class? Who is the spottiest boy?)

- creating fake accounts, hijacking or stealing online identities to embarrass a young person or cause trouble using their name

- sending explicit messages (also known as sexting)

- pressuring children into sending sexual images or engaging in sexual conversations.

Many parents think that cyberbullying is something that happens to other children, not theirs. Through the course of my work, I'm in contact with a large number of secondary schools, and every single one of those schools has had difficulties among their students relating to cyberbullying, sexting or exploitation of pupils. I know, having spoken to countless head teachers, that it occurs in every single year group and across both boys and girls. It is not a rare issue that parents can turn a blind eye to; it is happening everywhere. If your child hasn't experienced it personally, then they will know someone who has. The other week my son's primary school sent out a letter warning parents about a new app where users can make anonymous comments about other users and someone at the school had been targeted and cyberbullied using it. This goes to show that this issue affects primary-age children, not just adolescents, so parents have to address it. It's important to talk to your child about cyberbullying just as you would talk to them about bullying in the real world. Talk about things they might have seen or read online, and discuss how they could be hurtful. The important thing is to label it as bullying as, because it's behind the anonymity of a screen and not face to face, a lot of children don't see it that way. How will the other person feel when they read a post about them? Ask your child to put themselves in the other child's shoes. How would they feel if someone was saying the same thing about them?

What if your child is the cyberbully?

Because of the anonymity/safety of a computer screen, children (and many adults) think they can say what they want online, no matter how hurtful, and there will be no consequences. From an early age, though, we need to teach our children to stop and think before they post. Is what they are about to type something they would be comfortable saying to a friend in the playground? Would they be happy for a parent or their teacher to read the comments they are about to make?

I have a friend who tells her children that when they are online they should always imagine that she or their dad is behind them, reading what they are typing. If what they're typing is not suitable for their parents to see, then it's not the kind of thing they should be saying online. The same rules that apply to the real world apply online. You could also talk about the distinction between sharing or commenting on something because it's funny – and causing upset and hurt. One US study of 500 children aged eleven to fifteen discovered that 15% of them admitted to cyberbullying others.[5] Researchers found that girls were most likely to post mean comments, whereas boys posted hurtful pictures and videos.

If your child is accused of cyberbullying:

✪ Keep an open mind until you get the full facts. You are bound to feel angry, worried and ashamed, but don't let your feelings affect your judgement.

✪ Talk to your child calmly and get the facts. Let them tell you what happened. How many people were involved? How long has it been going on?

✪ Ask them why they posted or shared what they did. Was it because they thought it was funny or 'cool', or would get them extra likes? Or have they had a real-life disagreement with this person? Were they retaliating for something else?

✪ Don't take away their devices – they need to learn how to use social media and the internet safely and appropriately, so banning them from it won't solve anything, and could mean that it might happen again. (The only time I would recommend removing all devices is if your child has committed a very serious offence: then you can ban screen time or take away their device for a certain time, but there is no merit to banning screens permanently.)

✪ Cyberbullying means a child is not emotionally responsible enough to be online or on social media. Restrict their access or cut down the time they're allowed online. Anything they post has to be approved by you. Make it clear that if they prove themselves then eventually they can win back these privileges.

✪ Discuss how the person that the comments or the pictures were about might have felt. Try to make your child realise how powerful words can be.

Case study

Thirteen-year-old James was referred to me because his mood was very low – he was self-harming, anxious and refusing to go to school. His mental health had gradually been declining over a number of months, but he wouldn't open up to his parents about it. They were at a loss. He was reluctant to talk to anyone about how he was feeling and what had led to this. He needed urgent psychological therapy.

THE SESSIONS

It was clear to me that something traumatic had happened to James and there was something he was very, very afraid of – but it took several months of building up a relationship with him before he was ready to tell me what it was.

'I can't tell you because he'll come and get me,' he said.

It was only when I asked 'Who will come and get you?' that the whole horrific story came out. James told me that two years before, when he was eleven, he'd been a member of an online skateboarding forum. He'd been using his PC in his bedroom and his parents didn't really question what he did with his online time. James was in this forum one day when another child had begun chatting to him. This wasn't unusual, as James spoke to several children online. He told James he was an older teen and knew a lot about skateboarding, so James had been keen to message him. Over a period of months this person had begun to groom James – although James didn't realise what was happening. They started to private message each other, talking about skateboarding, passing on tips and sharing jokes. Eventually the teen had asked James for his mobile number so that he could text him and find out when he was going to be in the forum chatroom so he could log on at the same time too. One day, out of the blue, the boy FaceTimed him. To James's horror this 'teenager' was actually a man in his fifties. He threatened James and said if he told anyone about him then terrible things were going to happen to him and his family. He then got James to perform a sexual act for him. After that he FaceTimed James every few days and forced James to do the same thing. In their chats James had mentioned which school he went to, so the man would park outside James's school occasionally and wave to him. Understandably, James was absolutely terrified. This man had instilled such fear in him, and had told him that if he told anyone about what had happened he would kill James's family or tell everyone what he had done and shame him. He scared James into silence. He FaceTimed him at night and messaged him constantly during the day. James was in a permanent state of fear and panic. He had got more and more depressed and eventually he'd had a complete breakdown. His parents couldn't understand why – and James refused to talk to them.

INTERVENTION

When James disclosed this information, the man was still in contact with him. I explained to James that the first thing I had to do was to tell the police about him and what he had been doing. James was very fearful about this – he needed a lot of reassurance from me and the police that he was safe and this man couldn't hurt him any more or get to his family. At the same time, we also had to tell his parents what had happened – which James wanted me to do. Understandably, they were absolutely horrified that this had happened to their son and they hadn't known about it. James's mum in particular felt horrendously guilty that she had allowed this to happen and had not noticed her son was getting more and more depressed and withdrawn.

After what he'd been through, I knew his recovery would take a lot of work. I had to help James feel safe, and reassure him that the man had been arrested by the police and the abuse had stopped. I supported him throughout the whole legal process. Through therapy I helped him understand that what had happened to him wasn't his fault and nothing he had done had caused this. I also worked with his school to try to get him back into lessons. He'd had friends before but, when the abuse had started, he had retreated from the world. When a child has become disconnected from their peer group it's very hard for them to reintegrate again. It's like being in a race – everyone's trying to keep up, but you've stopped to have a rest. When you look around, there's no one around; you're on your own. Life really does move on quickly in that way. We had to work to help rebuild James's social links at school and help him feel confident enough to go out and spend time with friends. He was very fearful of anything to do with his previous favourite activity, skateboarding, so we arranged for his friends to take him back to the skate park when it was quiet. Gradually he reignited his passion for skateboarding.

OUTCOME

James's abuser was tracked down and was found to be grooming several other boys. He admitted his crimes and is currently in prison. James slowly started to attend school again, on a part-time basis at first then gradually built up his hours. He needed extra help due to all the time he had missed. He also managed to rebuild a couple of friendships.

He was so traumatised by what had happened and felt so unsafe about the internet that he didn't want to go anywhere near it. I said this probably wasn't realistic, so I helped him to face his fears and go back online. It wasn't a quick fix for James; it took months of therapy before his mental health started to improve.

Red flags

Are any of the following true of your child? They could indicate that your child has a problem with cyberbullying or grooming. If you suspect your child has been involved in cyberbullying – either as a protagonist or a victim – then you should alert your child's school, if the other children involved are at the same school. Today, schools should all have a policy on safe internet use, and should be able to help. Schools may also be able to offer counselling to children who have experienced serious bullying. You could also contact your GP for support from local health services or children's agencies.

- Your child seems anxious, tearful or low in mood. Or they have mood swings that seem very extreme.

- They are angry and agitated in a way they have not been before.

- ❶ They are reluctant to go to school.

- ❶ There is a change in their friendships: certain friends are not being mentioned as much.

- ❶ Their eating/sleeping patterns change.

- ❶ They are secretive about their screen time in a way they haven't been before. They are reluctant to use their computer/device in front of people.

- ❶ They are suddenly spending more time online.

Some solutions to keep your child safe on social media

- ➲ Teach your child to tell you straight away if they feel uncomfortable about anything that happens to them online. Actively encourage a policy of openness and speaking up.

- ➲ Use privacy settings. They are not fool-proof, but they are helpful. Learn how the privacy settings work on the social media sites your children use, and teach them how to control the information they make public or private. Make their Instagram and Facebook accounts private or protected so only people they approve can see their photographs and posts. Check their privacy settings regularly.

- ➲ As they get older, teach your child how to block or report someone.

- ➲ If a latency-age child does have social media accounts, then parents should always have access to them so they can keep an eye on what is being posted or commented on. Parents should know all of their child's passwords and follow them on social media in case anything slips through the net.

➲ Talk to your child before they go on social media and come up with your own rules that you feel comfortable with. Some parents insist that their child cannot post anything online without them approving it first, or their child can only log into their accounts on the parent's phone so they can closely monitor their activity.

➲ Talk to your child about online safety. A latency-age child should not be sharing their full name, their home or email address or the name of their school with people they do not know on social media.

➲ Children should only allow people they know in real life to have access to their social media accounts.

➲ Teach your child to think before they message or post something, whether that is commenting on someone else's post or putting up their own. Encourage them to think about whether they would say that to the person's face. If not, then they shouldn't say it in the virtual world.

➲ Encourage your children to tell you about good and bad experiences they have had on social media. Encourage them to come to you if they see or read something upsetting.

➲ Transfer 'stranger danger' into the online world: talk to them about grooming (in an age-appropriate way). Explain that on social media people are not always who they say they are. Do they know this person in the real world? Make it clear that they can talk to you about anything they feel uncomfortable about or don't know how to deal with. Parents should make sure that, wherever possible, latency-age children are using devices in shared spaces – so downstairs in living room/kitchen rather than bedrooms – so they can keep an eye on what their child is looking at.

➲ Control what your child is looking at. Set restrictions on either the device or the Wi-Fi so they cannot access inappropriate content.

Chapter 8

The online world for children with additional needs

Why children with additional needs are more vulnerable to the effects of the online world and might need more screen time, rather than less

If screen time is an issue for most parents then for parents of children with autistic spectrum disorder (ASD) it can be even more problematic, as ASD is associated with a strong tendency towards obsessions and, for many young children, particularly boys, this will centre on electronic devices, internet use and gaming. As a general rule, boys with autism tend to outnumber girls with autism by five to one. We are still not entirely sure why there is such a gender bias. We're unclear if autism genuinely affects more males than females, or if the assessment process is better at identifying boys rather than girls. It is also the case that girls tend to be better at masking their social problems. They can have superficial social skills which means we don't instantly spot that they are experiencing social difficulties.

One mother tearfully described to me recently how she was forced to lock all the family's digital devices away as it was the only way to control her autistic daughter's screen time. It had got to the point where she was watching the same music videos on YouTube day after day and playing the same online game. The mother had to find more and more inventive ways of hiding the devices, and her daughter was turning the house upside down in an attempt to find them. No one else in the house could use a device when she was awake, and she even started to attack visitors to try to get their phones if they used them in front of her. The mother was overwhelmed.

Thanks to my work, I know how digital devices can be a double-edged sword for children with additional needs. For many, the online world is their natural arena. The way that young people with ASD process information shows a tendency towards a more detailed, focused approach, and they are typically drawn towards patterns, repetition and sequencing. This can often mean that they have a strong aptitude for digital media and it is very appealing to them. Computer language, which the digital world is built on, is a binary system. It is logical, concrete and black and white. There are no grey areas. This way of processing or understanding information is synonymous with an autistic style of thinking, and the online world can be easier for children with autism to understand than the more unpredictable 'grey' areas of social/human interaction. For this reason it can be a haven for children with ASD, where their cognitive thinking gives them an advantage over peers without ASD. It's a place where they can excel, compared to the struggles they face in the 'real' world.

However, the downside to children finding refuge online is that they can be so drawn in that the 'real' world becomes a much less appealing place to be and they start having more of a virtual life than a 'real' one. When this tendency is combined with the natural obsessive element shown by many young children, then this can become a very real problem for parents.

Children with ADHD

Many children with ADHD are also drawn towards digital media. In particular, these tend to be gaming or online activities that offer a high level of multisensory stimulation such as noise, sound and movement. For this reason digital devices can be more stimulating or captivating than TV, and are a particular pull for ADHD children. Parents of children with ADHD regularly describe how the only time their children are still or are tuned in and focused is when they are on a digital device. This is because one of the core symptoms of ADHD is the inability to take control of your attention and direct it where you want it to go. People with ADHD are easily distracted and struggle to stay on task. We can all find it hard to focus on things, especially if they are not very interesting or stimulating. But for people with ADHD the threshold for what grabs their attention is much higher. Things must be loud and bright and visually entertaining to keep their attention. That's why computer games are so good at sustaining the focus of children, and adults, with ADHD.

Obsession with screen time

Although children with ASD are drawn to digital devices, this can cause problems because we are starting to realise that their additional needs mean they are particularly vulnerable to certain effects of screen time.

Researchers have found that children with ASD tend to have higher screen use than their peers. One study compared how children with ASD spent screen time, compared to their non-ASD siblings. Those with ASD spent over 60% more time playing video games and watching TV than all non-screen activities combined. They spent little time on social media sites. They also spent more time watching TV and playing video games than participating in physical or social activities, while their siblings spent more time engaged in non-screen activities.[1]

Research has also shown that children and teens with ASD have higher levels of problematic, or addictive, video game use. The same study also found that problematic video game use was significantly connected with inattention and oppositional behaviour. Boys with ASD who enjoyed role-playing games had higher levels of problematic game use and problematic behaviour, such as arguing, refusing to follow directions, and aggression.[2]

I have a practice specially working with children and young people with ASD, and I have become increasingly aware of the difficulties they are experiencing related to the use of digital devices.

Dysregulation and hyperarousal

The poor regulation of your emotional state is called *dysregulation*. If you have ever seen a young child on a screen getting very worked up then you will know exactly what this means. They might become angry, agitated, stressed or upset – perhaps the game they are playing is not going the way they wanted it to, or they are losing or have been killed. Or you might describe them as being 'hyper' – they are shouting and leaping about while on the digital device or immediately after they have come off it. Their body is having a physical reaction to this too – their heart rate and blood pressure are up, their adrenalin is pumping, and they are agitated and aggressive. It is almost like they are in argument mode with the device – yet at the same time they are too immersed in it to stop. This is how lots of children behave on devices, but children with ASD often have great difficulty in regulating their emotions, and so this effect can happen more quickly and more intensely. Being in a state of emotional dysregulation is not good for a child's physical or psychological well-being. It is not good for their body to be in this adrenalin-surging, fight or flight mode for long periods of time, and it's not good – from a psychological perspective – for them to experience psychological stress that makes them over-aroused, distressed and overwhelmed.

Children with ASD and ADHD often also have dysregulation in their dopamine system. High dopamine levels cause the brain to race, and overload it, and low dopamine levels affect attention and focus. Some clinicians say this explains the lure of screen time for these children – they are seeking a dopamine hit because their own dopamine system is sluggish.

One popular medication for ADHD children is Ritalin, which works by increasing dopamine levels.

Sleep problems

Around 50–80% of children with autism have significant problems falling asleep and staying asleep. They may also have low levels of melatonin – the hormone that helps to regulate sleep. One study found that prolonged video game use, as well as TVs and computers in the bedroom, can contribute to sleep problems in boys with autism. In addition, researchers found a significant association between the total number of hours spent playing video games each day and reduced sleep in the boys.[3] When working with children with ASD, my advice is to pay extra attention to develop a good bedtime routine and good sleep habits, and limit or avoid device usage before bedtime.

Increased risk of cyberbullying

Children on the autistic spectrum experience difficulties with judging social situations and understanding the subtle social nuances that other children of their age would be able to grasp. These difficulties can mean they are more at risk online as they are not able to 'read' a situation easily and could be open to exploitation or bullying. This means they may need extra parental guidance. One study found that children with special educational needs were sixteen times more likely to be the subjects of persistent cyberbullying.[4]

Some of the social-reading challenges children with ASD face in the real world also cause problems for them in the digital world. Many children with ASD talk to me about the difficulties they have in understanding text messages or talking to someone online because they don't have the additional information from facial cues and body language to help them understand what's being said. Also, some children with ASD feel bolder when they are not dealing with someone face to face and so say things via social media that they would not necessarily say in real life. I have also seen children with ASD falling into the trap of forgetting that online actions can have real-life consequences.

Children with ASD may also be particularly susceptible to online grooming, for the above reasons, so it is more important than ever for parents to monitor their child's screen time and know what they are doing online.

Screen time instead of social time

This is one of my biggest concerns for children with ASD. The virtual world gives them the chance to build social relationships and communicate with others in a way that they would find difficult face to face. In the digital world they can engage with another young person in a structured way around a shared activity (playing an online game, for example) and this allows them to know what to talk about and gives a focus to their conversation. Without this structure, they would struggle to manage the social interaction and to know what to say. For children who find it hard to make friends, screen time gives them a way to socialise and can really help boost their self-confidence and self-esteem. It can also give them an element of fitting in – it gives them something in common with their peers at school.

Problems can arise when devices become a replacement for the real world. Sadly, I have seen this with many children with ASD – the harder things get in the real world for them, the more they retreat into the virtual world, where they feel more secure. If children find school hard, can't make friends and find it difficult to socialise, then digital devices can become a refuge. Devices don't give you any feedback, you don't have to work out what they're trying to say to you – and that is a welcome relief.

This can present parents with a dilemma. There is a fine line between letting children do something they enjoy and naturally excel at, and making sure that devices don't become a fallback for everything. Don't let the digital replace the social. We cannot let children become too reliant on the virtual world, particularly in latency, when they are trying to build their social skills. Children with ASD need a lot of support and help when it comes to socialising, but avoiding it is not going to help them in the long run.

What can parents do?

It is important that parents are alert to the potential of screen time creeping more and more in to their child's life and act very quickly to manage it as soon as they notice it's becoming a problem – much more quickly than they might for a child without ASD. I also advise the parents I work with to ensure that their child balances their screen time with non-screen activities, including ones that offer them the opportunity to interact socially with others.

The importance of structure and a timetable

I emphasise to parents of children with ASD that they really do have to have a firm, structured plan in place. I remind parents to work with the way their child understands the world – children with ASD may take everything literally. With children without ASD it is possible to be a bit more fluid and tell them they can have two hours of screen time a day – it's up to them how they use it, but for a child with ASD you need to be much more specific and lay out exactly when their screen time is allowed and for how long. So, rather than saying they have two hours of screen time a day, which a child with ASD might find difficult to manage, give them designated time slots. A visual timetable would work well, as would a timer to show them when they have to come off, so they know exactly how long they have left. A 'now and then' board can be a useful tool too, as it shows them what they will be doing after they come off the device, so they can transition more easily to the next activity. A 'now and then' board is an A4 page divided in half. The left-hand side has 'now' at the top and the right-hand side has 'then' at top. Then you stick symbols or pictures to the paper to show what's happening 'now' and what will be next. For more able children you can write what's happening rather than use pictures.

It is crucial for parents to accept that screen time is likely to be very important to their child with ASD. This is sometimes the only area of their lives where they feel they are in control, and where they are skilful and successful. However, you need to make sure you have the right balance: remember, all the time they are gaming or looking at apps or YouTube, they are missing out on other developmental tasks that are important for all latency-age children.

More, not less, screen time?

Children with ASD have a tendency to get very interested in a particular subject and they will often have a special interest or a subject that they enjoy finding out about and talking about. I spend lots of time talking with children with ASD and now I am a lot more knowledgeable about dinosaurs, vehicles (particularly trains), insects, space, Lego and Star Wars. I remind parents that it is important to share these strong interests with their child. If there is something their child enjoys then help them to look up and gather information online. For a child with ASD, their strong interest is akin to having friends: it is something they invest heavily in and get a lot of pleasure from, and this should be recognised and supported by parents.

The benefits of screen time

The virtual world, if it is managed successfully, can provide many opportunities for latency-age children with ASD. (Children with ASD often perform better in tests when they are done electronically rather than with pen and paper.)

Researchers are trying to use this natural affinity that children with ASD have with screens to see if screens can be used to teach children social skills. Researchers at UC Davis Medical Center in California are trying to teach autistic children to make eye contact more readily by using virtual reality. The idea is that, by creating mock situations with faces appearing on computer screens, children can practise making eye contact in a situation they feel comfortable in.

Lots of apps are specifically designed to help children with ASD – from showing them appropriate behaviour in social situations and helping them learn emotions and facial expressions, to giving them prompts and visual timetables to help them manage their day.

Every child is different

Just like any parents, parents of children with ASD have to work out what is best for their child. I know from my clinical work that children with ASD can react differently to different digital media. Some children use digital devices to wind down, and find that time on a tablet can calm them, give their brain a rest and help them switch off from the real world. Other children with ASD may find any online time too stimulating. Watch your child on the device. Look at their posture: do they look tense, or relaxed, or are they still and focused? How easily can you get them off the digital device, and how is their behaviour afterwards? You will soon work out what works well for your child and what is too much.

I rarely advocate screen bans as I believe that most of the time they are unsustainable but, for children with ASD, I think parents need different and possibly tougher rules. It may be appropriate for parents to consider a partial ban of devices in certain situations or at certain times of the day to support their child's holistic development.

Case study

Seven-year-old Edward has had a diagnosis of autism since the age of four. He was very verbal and articulate, but struggled socially. Having previously been a very calm, rule-following child, he had been referred to me because his behaviour at home had suddenly got out of control. He was having temper tantrums, throwing himself on the ground and threatening to smash things around the house. His parents were struggling to deal with these outbursts. Some of that behaviour had crossed over into school and he was getting into fights with other children in the playground. His parents were at a loss to understand what was causing it.

THE SESSIONS

The first thing I had to do was to get to the bottom of what was triggering these tantrums. Edward's parents didn't know what was causing them, and it took a few weeks of probing to get an answer. Edward's parents mentioned that his favourite pastime was looking things up on his tablet. However, when I asked them what he was actually looking at, they didn't know. He wasn't gaming and he was quiet in his bedroom and they mentioned how he liked researching topics that interested him, so they had left him to it. When I asked Edward about his tablet time and what he liked to do in more detail he admitted he had developed an obsession with three 18-rated games with very violent content. He desperately wanted to play these games, but he knew his parents would say no. As he was very rule-abiding, he had got himself into a state about these games. He thought about them all the time and was getting very distracted at school. He would use his online time to research them, and had found some screenshots of the games and a small clip of footage which he watched obsessively. He was angry and frustrated: he had an internal battle going on and he didn't want to talk to his parents about it as he was scared his tablet would be taken away. Edward's mind was full of conflicting and upsetting thoughts about these games. This meant that he was on a 'short fuse' for much of the time and easily provoked into a fight or aggressive response.

INTERVENTION

I explained to Edward that we would need to tell his parents about how confused and upset he was, and how this was making him behave in ways that were getting him into trouble.

His parents were shocked, as they hadn't realised what he had been doing on his tablet. Rather than telling Edward these games were inappropriate (as I might tell a child without ASD), I knew that would cause him a great deal of distress and potentially cause more anger and disruptive behaviour. I explained to Edward's parents that we needed to work with his obsession. I talked him about what aspect of these games he was particularly interested in. It turned out he was fascinated by the game play, particularly the play that involved fighting and violence. He was interested in watching how the fighting played out between the characters, and also how the injuries were inflicted. Instead I got him to research things like *Horrible Histories* which was still gruesome and macabre, but was much more age appropriate. He wasn't a violent child – he was very calm and passive, but he had developed an intense interest in blood and gore.

I also allowed him, along with his mum, to do some background research on the tablet into these violent games – things like who had designed them, where the concept came from, how the animation was designed, when they came out, how many copies had been sold. He then made a digital scrapbook with all this information, which he then showed to me and to other people.

OUTCOME

Researching the games and creating a digital scrapbook seemed to help Edward's obsessive need. His behaviour very quickly calmed down, as he was no longer wrestling with his internal turmoil or feeling as though he was doing something wrong. As we allowed him to feed into this obsession, he felt much more able to manage it internally. Over time it slowly started to die down and he was able to move on – now he's interested in something completely different. It was his anxiety about what he was obsessed with that had been causing the problems, not the obsession itself. His parents also realised they needed to keep a closer eye on what he was doing on his tablet. They agreed he could only use it downstairs, where they were, and said that Edward needed to talk to them and show them what he was researching.

They made it clear that they were happy for him to have interests and that they would like to research things with him together in an age-appropriate way.

Red flags

Your child may have a problem if:

- They are obsessive about screen time to the point where they want to be on devices for hours at a time and refuse to do anything else.

- They are overstimulated by online time.

- Screen time leaves them exhausted or overly hyped up.

- Online time is taking over and gradually replacing all other aspects of their life.

Some solutions

- Managing this tendency towards obsessiveness can be very tough for parents. Parents of children with these additional needs need to be extra firm and extra consistent with their rules around online time.

- Try to vary what your child does online, and offer some digital alternatives. To stop them becoming obsessed with one particular game or activity, get them to play / do something different with their screen time. Mix up their digital routine.

- Accept that your child might need more screen time, not less, when it comes to researching any special interests.

- ⮑ Know who they are interacting with online (for a latency-age child, this should only be people they know in real life).

- ⮑ Talk to them about appropriate behaviour online and try to get them to recognise when someone is being cyberbullied. Encourage them to share anything that has upset them or made them feel uncomfortable.

- ⮑ Acknowledge how important the virtual world is to them. For many children with ASD, the online world is their natural arena. The ability to connect virtually offers them the chance to build social relationships and communicate with others in a way they might find difficult to do in real life.

Chapter 9

The link between excessive screen time and violence

Will playing video games make my child violent? How to help children understand the difference between fantasy and reality

A relative recently bought a virtual reality headset around to my house, and my children and I all had a go. I have played video games before and, to be honest, I've never been particularly interested in them, but I tried the headset out of curiosity and it blew me away. It immediately hooked me and I was surprised by how totally immersive it was. When I put the headset on, for all intents and purposes, I *was* Batman and I was fighting the baddies. It felt incredibly real. I also noted my eight-year-old son's reaction to it. He was immediately drawn in and wanted to know everything about it – how much it cost, where you could get it from, what games you could play on it. I could see he was considering asking for it for his birthday or Christmas.

I do welcome advances in technology and know that resistance would be futile – however, I do have reservations. One of my reservations about technological developments like VR is that latency-age children are not developmentally mature enough to distinguish between fantasy and reality. They become very absorbed incredibly quickly and I worry that, as technology develops and becomes more realistic, they are going to find it harder to disengage and switch back to the non-virtual world. If you apply VR to a first-person shooter game then the violence and gore will feel incredibly real and the line between fantasy and reality will be even more blurred for a child.

With my children, even though I've welcomed tech into our family life, I have always stood firm about not letting them play games containing violent content. I do let my latency-age son play games that are suitable for his age group that involve conflict, fighting and battles. I think parents need to educate themselves about the aggression and violence that is in games aimed at young children, and decide how comfortable they are about their child seeing it. Is their child able to handle what they see? I always say to parents: when in doubt, err on the side of caution. There is little to be gained by a young child being exposed to violent or aggressive content. It may lead to several problems, such as nightmares, psychological problems and unnecessary stress.

Age ratings on computer games, and why parents should stick to them

Computer games are rated by the PEGI (Pan European Game Information) system. A game's rating means it is suitable for players of that age or over – so a game rated 7 is suitable for children aged seven and over. There are five PEGI ratings – 3, 7, 12, 16 and 18. The 3 and 7 ratings are advisory only, so they are not legally binding for retailers, but the 12, 16 and 18 ratings are. It should also specify, next to the age range, why that game has been given that rating – for example, bad language or scenes of violence or sex. Parents need to be aware that there are also different age ratings systems for games played on phones, tablets or directly on the internet via a PC. For example, Minecraft is a 7 rating, but a spin-off called Minecraft Story Mode carries a 12-plus rating.

My professional view is that latency-age children should not be playing any game that is recommended for children older than they are, and parents should stick to these restrictions. They're there for a reason. I know through my clinical work that this is not happening. It is common for me to come across children aged eleven or even younger who play 16- and 18-rated games that contain graphic and brutal violence – games such as Call of Duty, Halo and Grand Theft Auto. This often occurs in families where there are older siblings in the house who play these games. Younger children are exposed to the games and this is harder for parents to control.

This means young children are seeing, processing and being emotionally affected by material that can cause them distress. Once they have seen something, they can't 'unsee' it. Parents need to think carefully about what they want their children to see, and know about, and to carry around in their minds on a day-to-day basis. I believe that young, busy minds have enough developmental work to do without the added demands of processing and managing potentially distressing images and information.

Some parents are quite happy for their primary-age child to play games designed for a much older teenager or adult. However, when I ask them if they would take their latency-age child to see an 18-rated film at the cinema then their answer is always no. When I ask parents why they let their child play an 18 game, their response is normally: 'Oh, it's only a game. It's not doing them any harm.' As adults, we know it is only a game and we can rationalise it – but a latency-age child cannot. The images in these games are incredibly realistic and powerful. Hopefully, they are not things a child will have encountered or experienced in the real world before, and they can be very alarming. Parents should ask themselves: do I want my child exposed to this level of virtual violence, or to become used to it?

A recent poll[1] showed two-thirds of parents whose children frequently played video games said they did not check restrictions, and 55% agreed they did not think that age restrictions mattered on games. I have come across countless children who have been traumatised by things they have seen in games that are not age appropriate for them. Most often these are violent or frightening images, rather than sexual. I recently worked with a seven-year-old boy who was back sleeping in his parents' bed because he had worked out how to disable the security settings on his device and had accessed and played what is marketed as a horror game. In this game a ghost follows you around, and you have to run away from it. The graphics are very good, it's very supernatural and creepy, and this poor seven-year-old was scared out of his wits.

It doesn't always have to be violence that causes trauma; it can be anything that plays on a child's fears. It might be that they're being chased, a creature is going to get them, or aliens are going to abduct them. Young children don't have the cognitive ability to understand these things are not real and don't exist in the real world. They're significantly more frightening for a latency-age child rather than a child aged over twelve, who knows that ghosts and aliens don't exist and who can rationalise the game and treat it as entertainment – whereas this little boy was genuinely frightened by what he had seen and was worried for his own safety. Sexual images, although they are inappropriate, are not usually frightening to a latency-age child. They're more likely to think they're a bit yucky. Sometimes if a child stumbles across sexual images or content, they might feel as if they've done something naughty or bad. Feelings of guilt are associated with sexual images, whereas violent content is likely to be associated with feelings of anxiety or panic. Many children who see violent images spiral into thinking they are not safe, that somebody could hurt them, or that bad people are coming to get them or their family. Sexual content leads to guilt, violent content leads to fear – and it tends to be fear that has the most immediate impact on children of this age. Children who have stumbled across inappropriate content online tend to be fearful and anxious. They have sleep problems, they wake up in the night or wet the bed. They also get separation anxiety: they don't want to be away from their parents, go to school or go to sleep at night. These are all signs of anxiety or that a child doesn't feel safe.

Does digital violence cause real-life violence?

Many parents want to know: can playing violent games lead to violent and aggressive behaviour? Violent video games have been blamed in a number of high-profile crimes, including the Sandy Hook massacre (when twenty-year-old Adam Lanza killed twenty-six people at a primary school in Connecticut, United States, in 2012). In the days after the mass shooting, much was made of Lanza's love of playing violent games such as Call of Duty, Combat Arms and Grand Theft Auto. I personally think a lot of other things come in to play when we are looking for reasons to explain something so incomprehensible – in this case, Lanza's mental health issues and his obsession with guns and the military. However, violent video games provided an easy scapegoat. No study has ever proved that playing violent video games causes someone to commit a violent crime – but much research seems to show a link between playing violent video games and aggression and lack of empathy. Some experts argue that studies showing negative effects are more likely to be published than those that do not, and that the research that exists is inconsistent and riddled with flaws. But I feel there are too many to ignore.

One analysis of 136 studies involving over 130,000 participants concluded that all studies showed that video game play is a causal risk factor for aggressive behaviour and for decreased empathy and prosocial behaviour (positive behaviours that assist others and show kindness and awareness of other people's needs and wishes).[2] Empathy is hugely important: we want our children to be able to recognise, and sympathise with, others' emotions and understand how they are feeling. This helps them know when they have made someone happy or upset. We also need our children to relate to other people's feelings: to put themselves in others' shoes so they can understand why people behave the way they do, and to help and assist other people because they can imagine how they would feel in the same situation.

It has also been suggested that playing violent video games can desensitise players to real-life violence. Researchers got participants to play a violent or a non-violent video game for 20 minutes and then watch a 10-minute film showing real-life violence. Their heart rate and perspiration were monitored. Those who had played the violent video game showed reduced physiological responses when watching the real-life violence. This suggests people who play violent video games get used to the violence and become desensitised to it.[3] Other studies have found that children who play violent games are generally more aggressive, are more likely to have been involved in physical fights, and get into arguments with teachers more frequently.[4]

However, one study showed that video gamers' aggression was linked to frustration, not the violence in a game. The study found that failure to master a game, getting stuck, or losing over and over again led to frustration and aggression, regardless of whether or not the game was violent. Researchers said such frustration is commonly known among gamers as 'rage-quitting'.[5] Another study suggested that it might be the competitive element of gaming that is responsible for the link between video games and aggression, not violence.[6]

Researchers have looked into whether pathological (problem) gaming can cause an increase in physical aggression. They found that time spent playing violent video games, rather than non-violent games, increased physical aggression and, interestingly, higher levels of pathological gaming, regardless of violent content, predicted an increase in physical aggression among boys. This study suggests it is the length of time a child plays games for, not necessarily the violence in them, that is linked to aggression.[7] Researchers believe that when boys spend a lot of time gaming, it starts to interfere with other important activities in their life such as school and homework. Their gaming starts to cause problems at school and with their parents – and when attempts are made to stop it, they get withdrawal symptoms, which leads to irritability and aggression. The study suggests that when a child is prevented from gaming, they experience withdrawal symptoms, which makes them aggressive.

Clearly, the research in this field is complex and there are many factors to take into account – but the overall message seems to be clear. Playing violent, or competitive, games for extended periods of time can result in increased aggression – do you want to risk this for your child?

Game transfer phenomenon

A recent study of over 1,600 gamers discovered something called *game transfer phenomena (GTP)*.[8] This is when players become so immersed in their gaming that when they stop playing, they transfer some of their gaming experiences to the real world. All of the gamers, who were aged between fifteen and twenty-one, had experienced some form of GTP, including being unable to stop thinking about the game, expecting something in the game to happen in real life, and confusing events that happened in video games with real-life events. The younger the player was and the length of the gaming sessions both influenced the severity of GTP experienced. I have met latency-age children who have described GTP to me, and parents who report that their children have acted out scenes from games. It stands to reason that if adolescents and young adults are vulnerable to the phenomenon of GTP, then younger minds will be much less equipped to handle, and prevent, it.

What should parents do?

As you can see from just some of the above research, video games and violence is a minefield – and a very confusing area for parents. With latency-age children, I would rather err on the side of caution when it comes to digital violence.

Violent games are about fantasy and escapism. Adults know that we can't really go round shooting people, stealing cars and killing zombies. These games involve some of the subconscious aggressive things that adults *might* like to do, and they are a safe place for us to express these fantasies. Young children do not have the same ability to understand the crucial difference between what we can do in the real world and what happens in the virtual world. What's entertaining and cathartic to us could be terrifying to them. As I mentioned previously, most parents would not dream of letting their latency-age child watch a violent 18-rated film. At least if you are watching a film then it is a passive experience: you are watching the characters on screen being violent. With a first-person shooter game, you, as the player, are carrying out the violence. These games expose children to the idea that violence is normal, fun and consequence-free; it is something that helps you win points or move on to the next level.

To expose these adult themes to children in latency is completely inappropriate. It presents children with cognitive dilemmas and challenges they are not equipped to solve.

We are holding them back from adolescence for a reason. We protect children from these themes until they are approaching adolescence, the purpose of which is to prepare children for adulthood. In adolescence, children begin to become aware of aspects of adult life, including sexuality and (sadly) violence. But in latency they are a long way off being ready to explore or understand these themes.

Some parents believe we should help our children grow up as quickly as possible. They argue that we need to equip them for the real world, and we can't shield them from reality and things like violence, otherwise it will make them naive and vulnerable and they will not be streetwise. I believe the opposite is true. We need to protect children from growing up too quickly. There is a strange paradox in our society where on one level parents are quite risk averse and want to protect their children and keep them indoors, whereas our grandparents would have left school and started work at fourteen. Yet in the virtual world we are exposing our children to things they are not ready for, emotionally or socially, and with which they can't cope.

With so much conflicting information available, I would rather take a conservative approach. If we are going to get this wrong it's better to get it wrong by being too cautious. I'm struggling – as both a psychologist and a parent – to think of reasons why we would want or need to expose latency-age children to digital violence. Even if we can prove it does not do them any harm, what is it adding? What is it giving our children? I certainly can't see any benefits.

How to help a latency-age child distinguish between fantasy and reality

✪ Encourage your children to express their feelings. Ask them how they feel when they see violence depicted in games or on TV – angry, sad, scared, excited?

✪ Help your children understand empathy for others. Ask them what would happen if what is depicted in a cartoon or a game happened in real life? How would they feel in real life if someone they knew was badly hurt?

✪ Remind them that real violence isn't a joke. Help them understand that when people get hurt, that is not entertainment. Talk about the difference between what happens in something like a cartoon compared to what would happen in real life.

✪ Compare video games and TV – would your child like to watch these things on TV? Children often understand that TV is acting and 'not real', but in computer games there are characters, not actors. Children can form a very strong attachment to characters; they don't view them in the same way as acting, which is like 'pretending' to them.

Case study

Ten year old Katie's parents brought her to see me. Having had no previous psychological problems, for the past three months she had been having difficulty getting to sleep and needed constant reassurance that family members were safe. She was also showing anxious behaviour, such as checking doors and windows at home, and continually asking if the alarm was on. She had always gone off to school fine, but was getting tearful about the idea of leaving her mum – which is unusual for a Year 6 child. Katie had been able to explain to her parents that she was worried about people getting into the house and hurting her or her family, but hadn't elaborated much more. Katie's parents had gone to great lengths to reassure her, but to no avail. They were getting increasingly worried and frustrated by her behaviour.

THE SESSIONS

Through conversations with Katie, I discovered that she had her own smartphone and enjoyed playing games and apps. It was all age-appropriate stuff such as Angry Birds and Pokémon Go.

Initially, I did not dig too deeply into her use of games or the internet, but as our work continued she alluded to more specific worries, and her issues with checking and sleeping. However, she still did not readily volunteer info about her online usage. After about six sessions she told me she had been playing a game one day and a pop-up appeared for another game. She had heard some of the boys in her class talking about this game, which was about zombies, so she clicked on the link. She knew it was an adult-rated game, but she was curious. The next thing she knew, she was playing a free trial run of this game where zombies try to catch you and eat you. Katie told me the graphics were high-quality and very realistic: at one point in the game, she opened a door and the body of a dead zombie fell out of the sky. It was clear by the way she described it that just a few minutes of playing this game had left Katie absolutely terrified. Her adrenalin levels were through the roof, her heart was thumping, and she was genuinely frightened. She started to think zombies were coming to get her and her family, and she couldn't get the terrifying image of the body falling from the sky out of her mind. Every time she went to sleep she would replay it. Katie was very concerned that she was going to be in trouble with her parents because she had played an adult-rated game, so she hadn't told anyone about it. Instead, she was trying to cope with this distressing experience on her own.

INTERVENTION

Unfortunately, Katie's worries are so common in young children I see. Adults know that zombies aren't real so there's nothing to worry about, and the game is designed to be as gory as possible. But this had breached Katie's fantasy/reality barrier and she couldn't believe the world was safe any more. It doesn't take very much for gentle, developing minds to be really affected by things they see online.

First, I needed to help Katie understand why she felt so anxious and fearful. We talked about the barrier between fantasy and reality. Latency-age children have good imaginations, so they believe in the tooth fairy and Santa and magical things such as Harry Potter. Even at the upper end of latency, they're not quite sure what's real and what isn't. Therefore I had to help Katie understand that this game wasn't real. We talked a lot about how games are designed this way, and I explained that she had been exposed to something she wasn't supposed to have seen at this age.

I also helped Katie talk openly with her parents about what happened, and worked with them to maximise the restrictions and safeguards on Katie's phone. Initially, Katie's parents wanted to remove her phone and take her 'offline'; I explained that this wouldn't deal with the issue or teach her any skills to cope with what had happened, making her potentially vulnerable to the same issue when she went online in the future.

OUTCOME

Slowly, I helped Katie build her resilience and understand that things that happen in games are not real. As she started to feel reassured that she and her family were safe, her anxiety gradually subsided. Unfortunately, this one small exposure had led to a significant problem.

Red flags

● Your child is re-enacting violent scenes from a game in everyday life.

● They have problems getting to sleep or staying asleep.

● They relate to other people as if they were characters from a game.

● They are secretive about their online time, and don't want you to see what they are doing.

● They show increased aggression – fighting, shouting, hitting, rage, emotional outbursts – which is new for them.

● They seem to be on a very short fuse and get angry quickly.

● They have new, excessive conflict with siblings.

● They show less empathy, not caring about the feelings of others.

Some solutions

➲ Stick to the age restrictions on games.

➲ Check exactly what your child is being exposed to. Are older siblings or friends' older siblings playing games designed for much older players that your child might be seeing/playing without your knowledge?

➲ Make it clear that behaviour copied from a game is not acceptable in real life.

➲ Start an open discussion about games and their content. Reassure your child that they can discuss anything with you, and they can tell you if they have seen something online that has upset or distressed them.

Chapter 10

How to stop screen time becoming a battleground

Safety restrictions, how to negotiate over screen time, and should screen time ever be used as a punishment or reward?

Most latency-age children love rules. They have gone from independent toddlers ('I do it') to starting school and understanding the importance of social hierarchy and following the rules. At this age they have quite a strongly developed sense of morality: it's all about doing things properly and following the rules to the maximum. So why, then, doesn't this rule-loving approach seem to extend to screen time?

Screen time is the issue that causes most arguments among all the families I come into contact with through my clinical work: they face a constant battle to get their children off their digital devices. Parents have described to me how their normally well-behaved children have tantrums, meltdowns, lash out at them and have even kicked and smashed a TV – all because they were asked to get off their screens. One mother described to me how, each night, she has to wrestle his tablet out of her eight-year-old son's hands, such is his determination to stay on it. One in four parents say they struggle to limit their children's screen time, and admit they find it easier to get their children to do homework, go to bed, or eat healthily than to turn off their digital devices.[1]

As well as the rule-loving part of latency, there is also a flip side. Children of this age can still get very consumed by the desire to follow their own agenda, much like they did when they were younger. If they can't get what they want, they may regress and have a tantrum.

Throughout this book I have talked about the idea of coming up with a timetable for screen time with your child. But what else can parents do to help control screen time – and avoid daily arguments?

Restrictions

Restrictions can be on the amount of time your child is on the device or on the content they are allowed to access – or both.

Restrictions on content

With a latency-age child I think it is always a good idea to restrict inappropriate content. You can do this by:

- using filtering software provided in most broadband packages
- using child-safe modes of websites, e.g. YouTube has a YouTube Kids app
- using child-friendly internet browsers such as Google's FamilyLink app
- setting restricted access on devices such as smartphones and e-readers
- using family security settings on your PC.

Restrictions on time

✪ Some games consoles have inbuilt timers so they switch off after a certain period of time.

✪ Restrictions on Wi-Fi: most providers will let you set individual times for connected devices, determining when and for how long they can have Wi-Fi.

✪ Some families have a 'charging station' and every family member leaves their digital devices there (preferably downstairs) to be charged overnight. This takes away the battle about devices in bedrooms. To make it fair, parents have to do this too.

✪ One family I came across had two different Wi-Fi routers – one for the children and one for the parents. The children's one turned off at 6.30pm every night. The children knew it was going to happen so there was never any argument.

✪ There are apps that help parents to restrict, control or time screen use, such as Time Lock, ScreenTime, ScreenLimit or OurPact.

✪ Using a timer, either on the device or your phone or a clock. This won't turn the device off, though, and it is useful if parents give children a five- or ten-minute warning before the timer goes off.

Don't rely on restrictions alone

Many parents use restrictions to limit screen time – but if you are relying solely on them, then are you abdicating responsibility? As the parent, you are the number-one regulator. You can use apps, timers and restrictions to back up your authority, but they should not be instead of you being in charge.

Parents need to take a dual-pronged approach. Technology regularly fails, children sometimes work out passwords, and, even though you may have restrictions at home, children use Wi-Fi in public places and go on their friends' devices, which might not have these filters. There are often loopholes in games or apps that parents don't know about. For example, I was recently talking to the parents of an eight-year-old boy who liked to play a very popular online game. It's a first-person shooter game, although it's rated 12 and the violence isn't very graphic and is quite cartoonish. His mum had looked at it and decided it was fine for him to play. He normally played it on his Xbox with his headphones on so the noise didn't bother anyone else in the house. However, one day he'd misplaced the headphones so the sound was on. His mum was upstairs when she heard the sound of adults shouting and swearing coming from the Xbox downstairs. When she asked her son about it, he explained that although they had lots of restrictions, there was a group chat function on the game that came on automatically when you played online with others – and he didn't know how to turn it off. His mum was absolutely horrified that her young son had been regularly exposed to this sort of language for months and she hadn't realised.

Parents need to talk to their children about how to recognise content that's not suitable for them and discuss what to do if they stumble across something online that makes them feel upset or uncomfortable. There are so many technological products and so much software available to block unsavoury content or kids' usage of the internet, but I believe the best way to teach children how to use the internet safely is through effective communication.

The London School of Economics' Media Policy Project found that some people favour time limits or the use of technical filters and software to monitor and restrict, while others prioritise 'enabling' or 'active' strategies including going online with your children (co-use) and talking with your children about what they do online. The project reported that parents who use a combination of approaches, modelling positive digital behaviours, and involving their children in setting limits have children who are more able to access the potential of, and manage the challenges presented by, digital media. Research has also found that taking a restrictive approach might avoid risks in the short term, but could limit children's digital opportunities in the long term, so it's preferable to build children's resilience instead.[?]

Sharing screen time

Sometimes understanding your child's screen time can help stop it being such a battle. Take an active interest in what they are doing on their device. It's important to acknowledge to your child that you know their online time is important to them; don't dismiss it just because you don't understand it. Take the time to ask them about their apps and games – and play some yourself. If you are worried about how much time your child spends playing Geometry Dash or Minecraft, and you struggle to get them off it, then get your child to show you how to play it.

Engage with them and ask them questions about the game – what do you like about it? What's the next level? Make them think about what they're playing and why they're playing it, rather than just mindlessly playing.

Communication is the key. If you listen and take an interest in what your child is doing with their screen time then you are building a relationship with them along their lines of interest and having a positive and engaged relationship with your child – which is vital when you need to be able instil discipline and rules. You want to be in a position where your child feels safe speaking to you openly about any online experiences that are upsetting or worrying without feeling you're going to take away their screen time. Learn how to engage with your child, not just police them.

Don't make screen time the Holy Grail

I believe that if you make screen time rare then it becomes even more attractive to a latency-age child and they will obsess about it. Parents often tell me that they ban all screen time for their children during the week, but their children are allowed it at weekends without restrictions. If this works for your family, then that's brilliant. However, I have found, from speaking to parents who do this, that it tends to lead to an 'all or nothing' approach. Parents find their children start using screens at 6am on Saturday, and for the rest of the weekend it is impossible to get them off it. I am not sure that abstinence/bingeing is the pattern we want to teach children. We want to teach them moderation, self-management and to feel they are in control of their own behaviour. Perhaps if they had 15 minutes or half an hour of screen time on a week night, it wouldn't become such a big issue?

I try to treat screen time as part of daily life. In my family we don't have 'screen time' as such – we have play time or free time. My eight-year-old son knows that once his homework is done then the hour before dinner is his to do what he wants with it – and this can include Lego, playing outside, watching TV or going on his tablet. It's up to him how he uses that time. Sometimes it includes screen time and sometimes it doesn't. It gives him choices and allows him to manage his own time, but with a reasonably clear structure. I do monitor what he does with this down-time: if he did use it for screen time every day, then I would talk to him about doing some other things to vary it. It's not a battle to get him off his device, because he knows the rules.

Screen time as a reward or a punishment

One so-called 'parenting hack' that went viral in 2016 was a photo of some wooden lolly sticks. On each one a parent had written a chore such as 'emptying the dishwasher' and next to it how much screen time doing that job earnt. Each chore equalled a different amount of screen time: the bigger the job, the more the child earnt. Some parents swear by making their children earn screen time by doing household chores or showing good behaviour. One poll of over 1,000 parents of children under eighteen found that 59% stop their children from using digital devices as a punishment for bad behaviour, but 51% say they should not be used as a reward for good behaviour.[3] Other parents have lists, and their child can't have access to screen time until everything on the list has been ticked off. Items on the list can include anything from homework and drawing a picture to writing a story and reading a book. Some parents like this idea of kids earning screen time and say it gives the child clear limits, yet also gives parents an element of control. Often, if the child doesn't use all the minutes they have earnt then they can carry them over until the next session. This is an issue that divides professionals and parents. Some argue that using screen time as a reward or an incentive sets it up as a goal for your child.

I think each family has to decide what works best for them. My son doesn't earn screen time, but I use a lot of what I call 'when' and 'then'. *When* you have finished your homework and read a book for 20 minutes *then* you can have screen time. I don't have an issue with children earning screen time if it works for you.

The other issue that frequently divides parents is the use of sanctions or punishments. In behavioural terms, there are only two key principles to follow. Reward the behaviour you want by giving it lots of attention. And ignore, or sanction, the behaviour you don't want. So should screen time, or rather the loss of it, be used as a sanction for undesired behaviour?

Again, this has to be a decision made by families based on what works for them. If you are struggling with your child's behaviour and one of the consequences of poor behaviour is a loss of screen time, that is completely understandable. It's very important to be consistent and reasonable. I strongly advise against using screen time as a reward and a sanction. It has to be one or the other. In a behavioural system, you should never lose the rewards you have earnt. This takes away all a child's motivation for working so hard for your reward if they could lose it again because of future behaviour. I think most of us would find this hard to manage. Your rewards are a reflection of your past efforts; no one can take them away, because they can't undo the good work that you have done.

Whichever rule you use, it's important that both parents are consistent.

Stay consistent

This is the number-one parenting rule: whatever your approach to screen time, however you choose to restrict or police it (or not), you must stay consistent. Intermittent reinforcement – sometimes saying yes and sometimes saying no – is a common mistake parents make, and it's guaranteed to result in increases in the behaviour you *don't* want. If you consistently say no to your child's demands for five or ten more minutes' screen time, then they eventually give up asking. They have learnt that you aren't going to say yes. If you sometimes say yes and sometimes say no, they will keep asking for more and more. They will think that they will eventually wear you down, because it has worked in the past.

If your child is showing non-compliant behaviour, this is often as a direct result of parenting inconsistency. Children will display behaviour they have been allowed to get away with. That is why consistency is key.

Case study

I was called in to see eight-year-old Lewis by his school because of a change in his behaviour over the past few months. He was answering teachers back, not following the rules, and being disruptive.

THE SESSIONS

As part of the process, I met with his mum and learnt she was having quite a few problems at home too. I went to see her, and she described to me how Lewis had lots of temper tantrums, particularly when he was asked to come off his tablet. He wanted to have access to his tablet most of the time, and his mum and dad had found it more and more difficult to say no to him because when they did his behaviour would become very aggressive and he would threaten to hurt himself or his younger sister or smash things. Dad worked away a lot, so Mum had been left mainly to cope with his behaviour alone and had become very frightened about what might happen. At the beginning she and her husband had tried to put some limits around Lewis's tablet use, but at times when he was very tired and couldn't fall asleep, his mum had let him have his tablet at bedtime as it made him tired and ultimately helped him fall asleep. Even though at one level she understood that having a tablet at bedtime was a bad idea, she had given in to him so she could have some peace and Lewis would settle and stay in his bedroom. Gradually, he began to nag for more and more time on his tablet, both before and after school. As his mum realised things were starting to slip out of control, she had started to say no a bit more and Lewis's behaviour had really escalated. He had been aggressive to his sister, and had threatened to smash the TV and Mum's mobile phone, and throw his tablet out of the window. His mum was ashamed and thought there was something wrong with the way she had parented him that he could be so aggressive towards her. She'd moved into a pattern where she conceded to his demands because that stopped the aggression. Sometimes she would try to be firmer and at other times she would give in. She was very inconsistent and unable to maintain her authority around this issue – and now his behaviour had spilled into school.

INTERVENTION

My intervention was primarily with Lewis's mum. We went back to basics to help her implement some boundaries and consistency. One of the things I had to help her realise was that it was OK for her to be firm and put limits in around Lewis's tablet – and that she *could* ride out his behaviour, no matter how much it escalated. I reassured her that when her husband was away she would have my support, and the support of her family (whom she had been keeping the problems from because she was so ashamed). One of the first things she did was to put some boundaries in place at weekends about when Lewis could and couldn't use his device. She got her father to come round, so she would say, 'Grandad's coming round – and when he's here you won't be on your device.' Grandad came round and tried to distract Lewis and also enforced his mum's rule. They told Lewis that if he didn't listen to them then they would turn the Wi-Fi off – or even the power, if they needed to. Grandad really backed up Lewis's mum and helped his mum feel that she had some authority behind her. Very quickly Lewis worked out that this was a coordinated approach and his mum was going to stick to it. He was told that if he broke his tablet all that would happen was that he wouldn't have a tablet any more – and the same with the TV. If he hurt his sister he would get very serious sanctions, and some of his favourite items would be removed. In this case we used the loss of his tablet as a punishment, as that was his predominant obsession. If he misbehaved, his mum increased the amount of time he had offline. We also drew up a screen time plan which she stuck to religiously: if things were going well, she stuck to the schedule. If things were going badly, she stuck to the schedule. The most important thing was that she was always consistent. She also worked with the school to enforce these boundaries and limits.

OUTCOME

Lewis gradually got the message that his mum was going to be consistent around the use of his tablet, and that becoming aggressive wasn't going to do any good. He still had a few outbursts of anger, but when he realised his mum wasn't going to change her mind he calmed down and begrudgingly accepted it, as it was preferable to no tablet time at all. Mum also spent some non-digital time with him, going to the park on their bikes, for example. She also started spending ten minutes at night reading him a story so he could fall asleep more easily without his tablet. Mum stuck to the timetable and got his tablet use down to a more manageable level. That consistent message was echoed at school and, as Lewis's behaviour started to improve at home, he also started to behave better at school as things were calmer at home.

Red flags

❶ Violent or aggressive behaviour when you ask your child to come off their device.

❶ Daily battles when you try to get your child off their device.

❶ Your child is on their device without your knowledge, or is disappearing with the device in the hope of avoiding detection.

Some solutions

⮕ Let them have screen time before an activity they're going to enjoy. So, rather than coming off a device to do chores or homework, have Lego out, a board game set up or a book open.

⮕ Before they get on the screen, talk to them about what they're going to do afterwards so they know the plan.

⮕ Make sure you stick to when they're supposed to come off. If you're caught up in work, Facebook or Twitter then you're more likely to give in to your child's demands of 'five more minutes', which then becomes ten and so on. Make sure you stick to the limits and are around to remind them to come off when they're supposed to.

⮕ If you use timers, then give then a five-minute warning before the timer goes off.

⮕ Stop your own screen time. If they have come off their device then, ideally, so should you.

⮕ Put firm boundaries in place, even if your child is distressed by having less screen time.

⮕ Be consistent, stay in charge, and retain control – your child will soon realise you're not going to change your mind.

Chapter 11

The good news – the benefits of screen time

Why all screen time is not equal and it's quality rather than quantity that counts

There is no doubt that technology has changed our lives and it is there to help and make life easier for us. I am very much of the view that we have to work with it and find healthy, balanced ways to integrate it into our homes and families so it doesn't become a problem.

There is no point demonising screen time and making it the enemy; that is not going to help anyone. It's important to be aware of the risks and stay in control, but I believe our job as parents is to help our children form healthy habits for the future. Our attitude towards technology shouldn't just be about protection and restriction. We need to think about the ways we can use technology constructively to get the best out of it, to bring our families together, and to promote connection, not isolation.

I know that screen time can give people an enormous amount of pleasure and fun. One of the most positive things is that our children have a whole world of information at their fingertips, allowing them to self-direct their own curiosity and learning. All of my children go off and research things they're interested in, and it always amazes me when they come back and tell me something they've found out.

We share the same meditation app so we can chart each other's progress and use the internet to research films for our family movie night. Digital devices allow me to communicate with my daughters and know where they are, and they will often FaceTime their brother when he is at his childminder, just to say hello.

Technology has also changed the way I deliver therapy. There are a huge amount of resources out there that can help the children I see – forums where they can talk to other young people and realise they're not alone, TED Talks, YouTube clips and websites. I recommend apps where a depressed child can rate their mood five or six times a day and visually see their ups and downs. Then we can track what's happening in their day to cause their changes in mood, which allows them to feel more in control. I certainly couldn't do my job as effectively today without access to the digital world.

Not all screen time is equal

When it comes to screen time, it's not just about quantity; it's about quality. There is a difference between your child using a digital device for two hours to Skype their grandparents, look things up on Google for a school project and make an animated movie, and using a device for two hours to watch videos on YouTube. Parents need to use their discretion and know what their children are doing on their screens. You would want your children to spend less time passively consuming digital content and more time on using screens constructively to create, learn and connect.

Passive screen time includes activities such as:

- watching TV

- looking at social media

- watching YouTube videos

- surfing the web.

There is absolutely still a place for passive screen time, though – if it is monitored. In small doses, it can be fun, relaxing and entertaining.

Active/creative screen time can include things such as:

- coding

- constructing a website or writing a blog

- making a film, animation or YouTube video

- taking photos and editing them

- learning a new skill

- making digital music

- playing computer games

- looking up information or researching something.

Technology in the classroom

The educational element of screen time has been much debated. However, technology is being used increasingly in the schools I visit.

The general consensus among professionals seems to be that digital devices in an educational setting can aid learning. Digital learning is interactive, it's visual, and children get instant feedback, so they know straight away whether an answer is right or wrong. Latency-age children, in particular, are more likely to take in information or understand a concept if a teacher can show and tell them about it. Interactive learning is not a replacement for traditional learning, but can work really well alongside it.

Two US studies that separately followed thirteen-year-olds and ten-year-olds who used tablets for learning in class and at home found that learning experiences improved across the board: 35% of thirteen-year-olds said they were more interested in their teachers' lessons or activities when they used their tablet, and the students exceeded teachers' academic expectations when using the devices.[1]

Digital devices can also help when it comes to reading. A recent study by the Literacy Trust found that technology offered a route into reading for disadvantaged children, and that children were more likely to enjoy reading more if they looked at stories using both books and a touchscreen, compared with using books alone.[2] Enjoying doing something (such as reading) is a key predictor of success for older children, so is this an indicator of the benefit of technology for young children?

Another study found that using tablets to do literacy activities in the classroom stimulated children's motivation and concentration. They offered opportunities for communication, collaborative interaction, independent learning, and for children to achieve more. Researchers also found that the noisier children were quiet because they were concentrating and the quieter children were using more language.[3]

My work involves regular classroom observations, and I have seen first-hand the rise of tech in the classroom. As a general rule this increase has been very positive and I have seen the level of engagement and enthusiasm that technology-supported learning can achieve – not to mention the enhanced benefits to children with additional needs of having their learning supported by technology.

The good side of gaming

Studies have shown there are several benefits to playing video games. Researchers have discovered that people who play video games are better able to process visual information and are more attuned to their surroundings. One study found that gamers who played action games had enhanced attention skills and made quicker, more accurate responses than non-gamers,[4] and repeated experiments have shown that playing fast-paced action video games can increase players' visuospatial ability. This means they are very good at working out shapes and patterns and the relationship between images.[5] Games can also promote social interaction and friendships in real life by giving children something to talk about with their peers in the playground.[6] Research has also shown that active video games can improve academic performance and reduce absenteeism, lateness and negative behaviours in the classroom.[7]

Researchers have also looked at the therapeutic benefits of playing video games. For example, they can be used to distract and relax children going through chemotherapy, and are also being used to help children with additional needs such as learning disabilities, ADHD and autism. An immersive VR programme now being offered on the NHS helps autistic children overcome their fears.

The child goes into a room surrounded by audio-visual images representing situations they might find overwhelming in the real world, such as getting on a busy bus or talking to an assistant in a shop. They move around the scenes with a tablet, supported by a psychologist. Research has shown that eight out of nine children who used this technique were able to tackle the situation they feared, and some were found to have completely overcome their fears even a year later.[8]

Positive ways for children to use screens

- ✪ Using the internet to research things, and looking up the answers to questions about the world.

- ✪ Learning how to code – using an app like Scratch Jr or a device like a Raspberry Pi to create their own computer games.

- ✪ Using educational apps to practise things like spellings and times tables.

- ✪ Starting a blog about something they are interested in, or an online diary.

- ✪ Using devices to take photos and edit them, or make their own movie.

- ✪ Using YouTube to learn a new skill, or starting their own YouTube channel (with a parent's permission and the parent monitoring and retaining control of the account) about something they are interested in.

Case studies

Eight-year-old Alice had suffered from anxiety and social problems in the past, and now she and her family were moving abroad for a year. She had made good progress and now had some secure friendships. However, her parents were concerned about how she would cope with the transition abroad and also whether she would settle back in the UK after a year away. I encouraged them, with her school's permission, to keep in touch with her old class while the family was abroad. Pupils FaceTimed her and sent her emails and photos, and she kept the class updated about her new life. Screen time helped her keep her social links, and so when she returned to the same school she settled back in very quickly. She felt she was still part of the class group and had maintained her friendships.

Eleven-year-old Emma had mental health problems. Going to school made her very stressed and anxious. Even with one-to-one attention she was finding the whole experience of going in to school and coping with the playground and lunchtimes too over-whelming. She wasn't coping, so we enrolled her in an online home-learning programme. This meant that every day during school hours she logged in to a virtual-learning classroom via her PC. It gave her access to teachers and a proper curriculum: even though she was not attending school, she was keeping up with her education. Emma did one academic year on the virtual learning programme while we helped her to slowly reintegrate back into her old school.

Chapter 12

Why parents should unplug too

Why parents need to practise what they preach – and how to cut down on your screen time

As a busy working mum, I'm constantly multitasking. The other night I was cooking dinner and checking emails at the same time as my daughter was having a conversation with me.

'Mum, please will you put your phone down and listen to me?' she asked me.

The thing was, I *was* listening, but it didn't feel like I was to my daughter. She didn't feel like I was paying her adequate attention – and she was completely right. Because I was multitasking, a bit of my brain was somewhere else at the same time and we weren't making eye contact. She certainly didn't feel listened to. Whatever other tasks I had to do, they were not more important than showing her I was available to listen to her. It was a sobering reminder that not only are we trying to manage our children's relationship with technology, we have also got to manage our own.

More and more I am struck by the irony that parents come to see me because they are worried about the amount of time their child is spending in front of a screen, yet when I see them in the waiting room, they are glued to their mobile phone or tablet. Even during a consultation, a parent's mobile will ring and they will automatically take the call. I have to tell more parents to put their phones away than children and young people I work with! You only have to look around you in a restaurant or a café, and you will see families staring at their devices rather than talking to each other. I find this highly concerning.

Even in child-focused places such as soft play areas or parks, which are full of families spending time together, the majority of parents are on their devices. When I began my career, mobile phones were not around. When I observed families out and about together, they were interacting or relating to each other. I'm certainly not saying that families were in a blissful device-free utopia where everyone was getting along and being happy, like a scene from *Mary Poppins*. But it was definitely different from today. Now I see families together – but apart. Each member of the family has their head down, concentrating on a device, not each other, and this change has been striking.

A recent study by Ofcom found that the average person checks their phone every 6.5 minutes: in the 16 hours most people are awake each day, they check their phones a whopping 150 times. According to Ofcom, we spend more time online than we do asleep.[1] Adults in the UK spend an average of 8 hours and 41 minutes a day on media devices, compared with the average night's sleep of 8 hours and 21 minutes.

Brits spend an estimated 62 million hours per day on social media[2] – the poll suggests that people spend around 34 million hours on Facebook each day, with a further 28 million hours on Twitter. Almost a third (30%) of the UK's 33 million Facebook users are on the site for at least an hour a day, with 13% spending at least two hours on Facebook each day.

Of the UK's estimated 26 million Twitter users, almost a third (31%) spend more than an hour a day on the site, while 14% – more than 3.6 million people – say their daily usage exceeds two hours. With these figures, it's no wonder that almost 70% of children think their parents spend too much time on their mobile phone, tablet or other digital device.[3]

So why is this a problem? I think there are a number of ways in which parental screen time can cause difficulties.

Children learn by copying their parents

We need to be aware that our approach to screen time has a direct impact on our children's screen time habits. If we model an over-dependence on screen time, then they are likely to develop exactly the same habits and difficulties. A latency-age child models and learns their behaviour from their parents. Their main model of how to behave at this age is still their family. Although outside influences like school and teachers are creeping in, children are still likely to default back to what their parents do and what happens in their family. Latency-age children still believe that all families are like theirs. Their benchmark for 'normal' is what they experience, or do, at home. It's very common for young latency-age children to be surprised when they go to a friend's house to play, to find that their friend's family does things differently to their family. They are much less aware of differences in society around them, and get their ideas and values about the world from what they see and experience in their immediate environment. This is very different to adolescence, during which children become much more influenced by their peer group. For teens, adolescence is a time to challenge parental norms and create their own. Adolescents live to be different, not to conform. They want to stand out (or stand away) from their family. Latency-age children are much less keen on being different. They like certainty, conformity, and fitting in. So they need their parents to be close at hand, keeping an eye on them and influencing them. They do not yet have the cognitive structures or maturity of adolescence to make sense of the world and develop their own ideas, values and behaviours.

A series of studies was carried out in 1961 by a psychologist called Albert Bandura to see whether children learn their behaviour through imitating and modelling what they see.[4] The experiment involved thirty-six boys and thirty-six girls, aged three to six. The children were put in a room with a series of toys as well as a giant inflatable clown (called a Bobo doll) that bounced back up again if it was pushed over. One group of children witnessed an adult in the room behaving aggressively with the Bobo doll. When it was their turn to play with Bobo, children who witnessed an adult pummelling the doll were likely to show aggression too. Just like the adult had done, the children kicked it, hit it and threw it in the air. They even came up with new ways to ill-treat it, such as throwing darts or aiming a toy gun at him. Children who were exposed to an adult who did not interact with the Bobo doll showed far less aggression towards it. As a result of his experiments, Bandura came up with a social learning theory, which tells us that children learn social behaviour through observation – watching the behaviour of another person. This was a revolutionary theory for its time, and has been key in helping us understand that a person's behaviour is shaped by their environment. Although psychology has moved on a great deal since Bandura and his Bobo doll experiment, his social learning theory is still relevant today – especially when you apply it to screen time. Children learn what to do from what they see around them. They learn through observing, imitating and modelling, and in latency-age children that is through the behaviour of their parents.

Being present for your children

Although technology has enhanced our lives in so many positive ways, it has blurred the boundaries between work life and family life/home life. Emails, Skype and smartphones mean we are always available, no matter where we are, and we are never truly 'off duty'. Modern life is all about multitasking, and it is rare to concentrate solely on one task. We live fast and we are constantly switched on. There is no down-time. We use technology to make life even more efficient and quicker. This means that we are not modelling (for our children) the importance of relationships, connections, taking the time to ponder or work things out. Then we wonder why our children are in a switched-on, continually wired state. As parents, we've forgotten what slow living looks like. Instead of slow parenting, we're multitasking and doing rapid parenting – and children are seeing this and thinking this is normal life. We are constantly trying to justify our online behaviour to them: 'It's for work', 'Give me five minutes', 'Hang on' are all things I hear parents (including me) telling their children as they check their phone, laptop or tablet. Unfortunately, a latency-age child doesn't really understand. All they see is their parent interacting with a screen, not them.

The head teacher of one primary school recently put up signs asking parents when they collected their children at the end of the day to 'greet your child with a smile, not a mobile phone'. Many children are used to seeing the tops of their parents' heads all the time, and they are getting the message that their parent's device is more important than they are. Current latency-age children have been born into a digital world, and many don't know what their parents look like without a device attached to them. Smartphones are almost like an extension of our hands.

Latency-age children need that one-to-one connection. They need to know they are important and they have your full attention. They need you to put your phone away when they are trying to share important news with you about school or their friends. They need eye contact and to know you are really listening, not distracted by the ping of a text message or a tweet. Most of all, they need you to be present.

This is because latency-age children have less ability to emotionally regulate and process the world than older children. One of the skills they need to develop in order to do this is to understand their emotions and feelings – namely, their mental state. Being able to understand your own mental state and the mental states of others is called *mentalisation*. It is a vital skill, and one that young children need to develop to learn how to process and manage their own feelings and to be able to relate to and empathise with others.

Parents have a crucial role to play in helping children to develop their ability to mentalise. When a parent helps their child understand what they are feeling, by labelling it or describing it for them, this develops their mentalising skills. We can all relate to seeing a distressed toddler who is overstimulated and having a meltdown. Their parent may say, 'I think you're tired and need a nap.' The toddler is adamant they are not tired, but they are unable to calm down. The parent takes their child's feeling of distress and explains it and manages it for them. When a parent says, 'I am feeling cross with you because of your behaviour' or 'I am feeling very proud of you because . . .', it helps to develop a child's mentalising capacity. However, if your mind is occupied by something else, you are unable to help your child mentalise and offer a reflective space for them – which is a big problem.

We have known for many years of the negative effects on children of not having had good enough mentalising support from their parents. Children who have suffered emotional neglect are at significant risk of future emotional and mental health difficulties. I believe we are now facing a generation of children who are growing up receiving less mentalising support from their parents – not because they are unwell or unable to do so, but because they are unavailable due to concentrating on their digital devices. Either the parents are disengaged from their children or the children are disengaged from their parents.

I call this the 'heads-down generation'. Mentalising is about talking things through with your child, talking about their day and what happened, and helping them make sense of it and process it. I believe this is especially important with young girls, who often need lots of help to understand friendship groups and the accompanying dramas. If you're not talking to your children, then that is a massive problem.

I often say to my own children that if they are stuck with a problem and they have maxed out their problem-solving resources then they should come and ask me because I've got different problem-solving resources, and maybe if we put our resources together, we can work something out. If children don't think they can approach their parents, then they are left to think things through on their own, and they can get very stuck. If a young child is keeping their problems to themselves then they are likely to get anxious and distressed. As the saying goes, a problem shared is a problem halved. If you're a child that is more important than ever. But children see their parents on their devices and think they can't bother them. A family can sit and watch a TV programme together and talk to each other while the programme's on, whereas I think you would be less likely to try to chat to someone who was watching a tablet on their own because it feels more like interrupting.

Because of our reliance on technology, we're losing many of the everyday opportunities we had previously to talk to our children. For example, in the car. A few years ago, a car journey would have been an opportunity for children and their parents to chat; now it's more likely that the children are in the back glued to their tablet, iPod or phone. I know toddlers who can't even go on a ten-minute car journey without being entertained by a tablet. We're losing lots of opportunities to talk to our children and help them make sense of the world. We parents have to make space in our day for these important, everyday moments. Even five or ten minutes make a difference – such as chatting to your child while they're in the bath or just before they go to bed.

In a video for Start-Rite's 'Pass It On' campaign, children aged seven to eleven were asked how they felt about their parents' use of technology. Most children said their parents were obsessed with technology, which made the child feel lonely, stressed or ignored.[5] It's pretty clear that, whatever we feel about our devices, our children are telling us that our behaviour is making them feel unhappy – and that our devices are getting in the way of our parenting.

Practising what you preach

Parents need to model the kind of behaviour around screen time that we want our children to practise. It's not just about making a screen time plan or timetable for your child; it's about making one for your entire family. You can't tell your child to get off their device if you're checking your emails while you're saying it. You have to model what you expect them to do. If your mobile rings, rather than leaping up and grabbing it, say, 'I'll let that go to voicemail, because I'm talking to you.' These actions send our children the message that they are more important than screens, and that you (and they) prioritise face-to-face interaction and connections over being a slave to technology. If we want our children to have a healthy attitude to screen time, then adults must have one too.

Earlier in this book I advised parents to take devices out of their children's bedroom at night, but how many parents sleep with their phone next to them? How many of us check our phones first thing in the morning and last thing at night? I know adults who wake in the night and immediately log on. Many adults have bad sleep habits or insomnia as a direct result of misusing their phones.

Our children have grown up with screens, tablets and smartphones. They don't remember a time without Facebook or YouTube. Parents, on the other hand, do, but we are now battling with our own obsession with technology and digital social interaction. If we can't model switching off, being calm and just engaging with our family then what hope is there for our children?

We know that latency is a crucial time to develop social relationships, to learn how to interact and connect with people. But many families are not sitting down and talking to each other; everyone is in their own world, busy on their devices. Families have changed because of the online world. Today, people are more separate. Families very rarely sit around the TV together any more. Parents have to seek children out and children have to seek parents out. It's very easy for parents to think that because they're physically in the same house as their children then they're spending time with them, but that's often not the case.

I know this from observing my own family, and families I visit as part of my clinical work. I go into a family's house, and my first job is to get them together in one room so I can meet them all. And this usually involves someone texting everyone to tell them to come downstairs and meet me. We are all present – but not connected. We are there – but we are not interacting. People are scattered around by themselves. I see that in my house. Dinner's finished, we've tidied up and suddenly everyone disappears. They retreat to different rooms and do their own thing (mainly involving screens). I think parents need to be aware of this disappearing act. I know that we all work long hours and not all families have the opportunity to eat together, but if in one day all the members of your family haven't been in the same room together even once, and this happens frequently, then I suspect you're not having enough family time. If you regularly feel that you don't have a proper face-to-face conversation with anyone in your family, that should worry you.

What families can do together

I find that many families today need help to know what to do with their children – they have forgotten how to spend time with them.

Parents will often say they can't afford to do anything with their kids, but you don't need money for many activities. My colleagues and I have compiled a list of twenty or thirty free or very cheap activities for parents to do with their children.

I always recommend that parents should try to spend ten minutes playing one to one with younger children every day. It should be child-led play – so that doesn't mean getting a game or a puzzle out. It should be about picking some toys and then letting your child take the lead. The parent should narrate the play: if your child picks up a dinosaur, you say something like, 'Oh, you've got the dinosaur. I wonder where he's going?' and allow your child to tell you. It may feel a bit false at first, but there's something really powerful about joining in with a child's play and letting them lead it. This can work really well even with children up to the age of eight. It's simple, but it can make a massive difference to your child. Spending time each day to do something with your child where you're focused on them is hugely important – and everyone can spare ten minutes. If you have a slightly older child, then you could draw with them or make something with them.

Other activities families can do together include board games like Twister, Articulate or Monopoly. I think doing a jigsaw puzzle together is fun, and you can often pick up a 250-piece puzzle in a charity shop for a few pounds. You could do crafts (papier-mâché, model making) or science experiments with things you've got in the house. Look at www.activityvillage.co.uk and theimaginationtree.com for inspiration. Also, YouTube and Pinterest are good sources of ideas for science experiments and crafts to do at home.

How about potato printing, painting or making a collage? You could make slime by mixing cornflour with water. Do origami (www. origami-fun.com) or make paper aeroplanes, then have a competition to see whose plane flies furthest. You could make a marble run out of tin foil and empty cardboard tubes from kitchen roll, or set up a domino run. There are so many simple, cheap and fun things parents can do with their children to spend important bonding time together.

Don't forget the benefits of getting fresh air and exercise too. Take your children pond dipping, birdwatching, den building, cloud watching, or go out for a walk and ask them to spot wildflowers of every colour. There are lots of country parks that are free to enter, and lakes, rivers and canals to walk beside. The National Trust has a list of things for children to do before they're 11¾ (www.nationaltrust.org.uk/50-things-to-do) – how many of these have you done?

In my family we go out for a walk every day. We have a dog, and I'm a massive fan of everyone getting their coat on, whatever the weather, and going out. Even reluctant members of the family can trudge through the woods for ten minutes. My son will climb trees and my daughter will pick flowers that she can press, and we all chat. When you go out for a walk, you can talk to your children without making direct eye contact with them, as you walk along side by side, and children, particularly teenagers, often find it easier to talk like this.

My children can confirm that cooking is not at the top of my leader board of skills, but I can always rustle up a batch of brownies or fairy cakes from memory, and that is an excellent way to engage even the most resistant child. We can even look up the instructions on YouTube if we get stuck.

How to cut down on your screen time

Even though I have written this book about the dangers of too much screen time, I'm as guilty of overusing screens as the next parent. I talk to my children about their screen use all the time, and how devices shouldn't take over their lives – and now they tell me off about it too.

Parents need to be stricter with ourselves. We must cut down our own screen time and reset the boundaries about acceptable screen use around our children. It's all very well saying your child has to come off their tablet and connect with you, but that's not going to work if you're glued to your device and not available to connect with your child!

One study by researchers from the University of Washington[6] surveyed 249 families with children aged between ten and seventeen about which rules around technology were most important to them. When researchers asked the children what technology rules they wished their parents would follow, the answers included:

- ✪ Be present – children felt their parents shouldn't use technology at all in certain situations, such as when a child is trying to talk to a parent.

- ✪ Moderate use – parents should use technology in moderation and in balance with other activities.

- ✪ Not while driving – parents should not text while driving or sitting at traffic lights.

- ✪ No hypocrisy – parents should practise what they preach, such as staying off the internet at mealtimes.

Children also said they found it easier to follow household technology rules when families had developed them collectively and when parents lived by them as well. None of this should come as a great surprise to any of us. What surprises and worries me is how hard it is to get families and parents to follow this very sensible advice.

A lot of screen use is about habit. We constantly check our smartphones and devices without being mindful about it. How many of us have been sucked down the rabbit hole of screen time? You check one tweet or text and, before you know it, an hour has passed. Devices mean we are constantly multitasking, but multitasking doesn't work; it makes us less efficient. Devices like tablets and smartphones are having a negative effect on our sleep and our relationships. We text rather than ring people, and even when we see people face to face we are not fully present as often we have one eye on our mobile phones.

Screen time has also led to a loss of focus; we are constantly interrupted by every beep and ping of our smartphones and if we don't check them then we worry that we are missing out. We have all heard of FOMO – Fear Of Missing Out. In fact, the term was added to the *Oxford English Dictionary* in 2013. It's the idea that your peers are doing something better than you, that you're not part of. Constantly clicking on social media can lead to a bad case of FOMO.

Many of us also feel anxious if we're away from our phones. There is now a name for this fear: nomophobia – the fear of being without access to a working mobile phone (literally a combination of 'no', 'mobile' and 'phobia'). As adults, we have to realise that it's not only our children who are being affected by screen time; we are too.

Log your own technology use

Record how much screen time you're actually having. Write down every time you check your phone, tablet or PC and for how long. There are a number of apps, such as Moment (iPhone), BreakFree and QualityTime (Android), that will record how often you check your phone, how much time you're spending on it, and which apps are taking up most of your time. Once you know how much time you are spending (or wasting!) on your devices, you can make a conscious effort to cut down.

Limit your non-work screen time

Decide how long you want to spend online and then set an alarm on your phone to signal when that time is up. Or use apps to help you block distractions on your devices and divide up your time. App and website blockers such as Freedom and SelfControl allow you to come up with your own schedule so you can decide what you want to block and for how long. Some apps allow you to block the whole internet, and StayFocusd restricts the time you can spend on certain websites or apps. Once you use up your allotted time, then the site is blocked for the rest of the day.

Turn off notifications

Do you really need to know every time a friend posts something on Facebook or someone comments on your Instagram? Endless beeps and vibrations will constantly interrupt you, encourage you to check your phone, and will stop you from being able to focus on anything else.

Out of sight, out of mind

Put your device in a different room or at the other side of the room or put it away in a drawer. You are less likely to keep checking it if you can't see it or if you have to get up to do it.

Change the settings

Turn your phone to silent, airplane mode or Do Not Disturb. Silent mode means your phone won't ring, but it will still vibrate when you receive a text or call and the screen will light up. The Do Not Disturb (DND) option on an iPhone stops notifications, alerts and calls from making any noise. Your phone is still connected and will receive calls and data, but the notifications are turned off. You can customise this, so even when your phone is in DND mode you can still receive calls from certain numbers – such as a babysitter or your child's school, for example. In airplane mode, your phone isn't connected to Wi-Fi or any network and no one can call you.

Have an app audit

Regularly go through your apps and delete any that are not essential. Decide what you actually need (for example, Google Maps, Uber, banking apps) and what you can live without (games, social media, news?). Do you really need to have Facebook and Twitter on your phone as well as your laptop? Regularly change the position the apps are in on your phone – switch them around so you don't get into the habit of automatically clicking on the same apps without thinking about what you're doing. Moving them is more likely to make you stop and think about what you are looking at.

Buy a watch and an alarm clock

If you have an alarm and a watch, you won't need to constantly look at your phone to check the time or keep your phone in the bedroom to wake you in the morning. Turn off your devices at least an hour before bed and ban phones from the bedroom so you won't be tempted to check them during the night or first thing in the morning. Try to get up, dressed and have breakfast before you check your device, so you can ease yourself into the day and wake up gradually.

Allow yourself to be bored

Instead of automatically reaching for your phone when you are on a bus or train, or simply waiting for an appointment, allow yourself to sit and be in the moment. Look around you or out of the window, or read a book or newspaper instead. Notice what's going on around you – or talk to people. Today we rarely allow ourselves to be bored and have any mental down-time away from a screen. Some experts believe we have our most original and creative ideas when we allow ourselves to be bored, without constant stimulation.

Go back to basics

If you are really struggling with your smartphone use and finding it impossible not to check it, then swap your phone for a basic mobile phone without internet access that you use just for texts and calls. Nokia recently announced that it is relaunching its 3310 model in response to consumer demand for simple mobiles.

The upside is these phones are much cheaper to buy and run, and less of a security risk if they're lost or stolen (think of all the information you have on a smartphone).

Set an online schedule

In earlier chapters, I talked about making a plan for your child's screen time, so do one for yourself too. Only check social media or emails at certain times of the day. Reply to text messages in batches two or three times a day (unless they're urgent) – this is much more efficient and less disruptive than replying instantly as you receive them.

Do something without your phone

Try going out for short periods without your phone. Go to a yoga class or for a walk.

Swap the virtual world for the real world

Ban yourself from looking at your phone whenever you are interacting with another person, whether you're out for a meal with a friend or at the supermarket checkout. Instead of sending a text, make an effort to meet up with a friend in real life or phone them for a chat. If you are talking to someone, regardless of who that person is, and your phone rings, don't answer it.

Make designated screen-free times and areas of your home

Ban devices from your bedroom and even the toilet. My son often takes his tablet into the loo, which drives me mad! Come up with some house rules that everyone has to abide by, such as no devices at mealtimes or bedtimes and during your child's bath and story time.

Lock your device with a long, complicated password

Choose a password you can remember, but which is long (at least twenty characters) and complicated to type, and which includes numbers and symbols. Entering it every time to unlock your phone will take a while and you'll get frustrated, which should put you off checking it so frequently.

Breathe through the urge

If the urge to check your device is overwhelming then breathe through it. Close your eyes and take a deep breath. Inhale for three seconds and exhale for three seconds. If the urge is still there, then repeat the breathing.

Case study

Lisa was a single mum with two daughters, aged seven and five. She had contacted me because she was having difficulties with the children's behaviour. She was anxious because the girls were fighting, they were disruptive and hard to manage, and she was worried they had ADHD.

THE SESSIONS

I went to meet Lisa at home and get some information about the family's everyday life. I met the children and, although I didn't have any concerns about their development, their behaviour was certainly challenging. They refused to get ready for school in the mornings, they wouldn't behave at the dinner table, and they were particularly disruptive at bedtime, when they kept getting up. Lisa said that she found it very stressful because she had recently started working from home and did a lot of work in the evenings. Now she wasn't working in an office, she had decided to take the girls out of the childminder's after school and look after them at home instead.

My main feeling on meeting the girls was that they were fighting and misbehaving in order to get their mum's attention. As is very common in young children, getting some attention, even negative attention, is better than no attention. Over the course of several home visits, our conversations were continually interrupted by Lisa's mobile ringing or Lisa stopping to answer work emails. I noticed that, when I tried to set up appointments with her, she found it hard to set a time to meet me face to face, yet she responded very quickly to texts or emails. My feeling, as a therapist, was that it was really hard to get hold of her and for her to be available. If it was difficult for me, them how difficult was it for her children to feel they had her full attention?

INTERVENTION

It had become part of normal life to her, but Lisa was unaware of how work and her digital devices had intruded into her life and were making her children feel neglected. Understandably, she was very upset and felt very guilty about this when we talked it through. I could see she wanted to do her best for her children as well as provide financially for them. I explained that she might be physically present, but her daughters felt that she was not there for them. At the childminder's they had been given lots of attention: however, although now they were at home with their mum after school, they weren't getting nearly enough attention, and that is why they had started to misbehave. Lisa generally put the TV on for them while she was working on her laptop.

We looked at simple ways Lisa could be more available for her children, such as setting clearer boundaries between work and home. We discussed the idea of the girls going back to the child-minder two days a week after school, giving Lisa two long days to work. Then, on the days the girls were at home, Lisa could turn her phone to silent and close her laptop for a couple of hours after school. We also talked about Lisa giving each of the girls individual attention, as well as learning how to play with them again. Their behavioural problems at bedtime were because they were trying to get Lisa's attention: they were competing with her laptop and phone, and always losing out.

Lisa started giving each of the girls their own individual bedtime routine. They each had a bath and then she would read them a story. It gave her time to chat to them about their day, and help them process anything they were struggling with.

OUTCOME

Within a few weeks of making these small changes, Lisa noticed her daughters' behaviour slowly started to improve. Having two separate bedtime routines took longer, but it saved her time in the long run, as the girls were much more settled at night. Lisa scheduled her time more strictly in the school day so she felt she had more time in the early evening – she was able to switch off for a couple of hours and spend time with her daughters when they got home from school. Then she could do more work, if she needed to, after they had gone to bed. She also started putting her phone on silent and keeping it out of view so she wasn't tempted to look at it. These were all very simple changes that made her girls feel as though Lisa was there for them without them having to compete with a digital device. As a result, Lisa felt calmer and more in control. She also felt less stressed at trying to juggle work and her children.

Red flags

- Your child often comments on, or challenges you about, your screen time.

- You feel anxious when you are away from your phone/device or if you have a low battery or are out of Wi-Fi range.

- You feel guilty about how often your child finds you on your device when you should be interacting with them.

- You can't have a conversation without checking your device during it.

- You leave the dinner table when you hear your phone beep to check a text message, or you text while at the dinner table.

- You are on your digital device while you talk to your partner or children, so you are never fully concentrating on them.

- You check your phone first thing in the morning and last thing at night.

- You sometimes check your phone in the middle of the night.

- You answer your phone if it rings when you are in the middle of a conversation with your child.

- You are regularly on your phone when you are with your children, including when you are out doing things together.

- You find it hard to sit and wait for something without reaching for your phone to fill the time.

Some solutions

- Lead by example. You need to model the sort of behaviour you want your child to copy. If your head is always buried in a device, why shouldn't your children do the same?

- Set rules around your online time as well as your children's. Talk to your child about what the rules should be. Perhaps you will promise not to check your emails or take calls when you are putting them to bed. Ask them what rules they would like to see for your digital time.

- Set guidelines that apply to the whole family: for example, no devices at the dinner table, no devices after a certain point each evening, or designated screen-free time.

⮕ Create good bedtime habits for yourself as well as your children. Try not to look at your phone before bed – perhaps leave it in a different room to charge or put it in a drawer. You will be less tempted to check it if it's not in front of you.

⮕ Allocate certain times of the day to check social media so it doesn't interfere with family time.

⮕ Remove your phone from your bedroom at night. Buy an alarm clock instead.

⮕ Do a digital detox for a day: see what it feels like to not be at the mercy of your phone.

⮕ Switch off all non-important notifications. Do you really need your phone to beep every time a friend posts something on Facebook?

Acknowledgements

My first 'thank you' must go to all the families and young people I have worked with. They have made me into a psychologist, and I learn something new from every young person I have the honour of working with.

To the Lot-Liv-Iz trio, who are the reason I do what I do and I am who I am. Thank you for being my eternal guinea pigs and having patience and wisdom beyond your years (and mine).

Thanks to my literary agents, Rowan Lawton and Rory Scarfe, for all their help and brilliance; to my publisher Lindsey Evans for commissioning this book and making this idea become a reality; and to Heather Bishop for her sheer graft in making it happen.

Finally, to some very long-suffering friends and family – who deserve more than a mention. You are the people who keep the wheels on the wagon and without whom not much would be possible. To Debbie, Elaine, Holly, Karen, Rupert, Victoria C., Victoria D. and Ollie.

Notes

Introduction

1. Ofcom (2016) 'Children and parents: Media use and attitudes report.' Available at: https://www.ofcom.org.uk/research-and-data/media-literacy-research/children/children-parents-nov16.
2. Channel 4 News (2015) Parents Poll. Available at: http://www.comresglobal.com/wp content/uploads/2015/09/Channel-4-News_Parents-Screen-use-Survey_September-2015.pdf.
3. Dr Aric Sigman, quoted in the *Huffington Post*, 22 May 2012. Available at: http://www.huffingtonpost.co.uk/2012/05/21/parenting-tv-time-bad-health-children_n_1533244.html.
4. Public Health England, quoted in the *Daily Telegraph*, 16 May 2014. Available at: http://www.telegraph.co.uk/news/politics/10835157/Too-much-time-on-web-gives-children-mental-health-problems.html.
5. Kabali, H.K., Irigoyen, M.M., Nunez-Davis, R. et al. (2015) 'Exposure and use of mobile media devices by young children.' *Pediatrics*, 2015–51. Available at: http://pediatrics.aappublications.org/content/early/2015/10/28/peds.2015-2151.

6. Corder, K., Atkin, A., Bamber, D. et al. (2015) 'Revising on the run or studying on the sofa: Prospective associations between physical activity, sedentary behaviour, and exam results in British adolescents.' *International Journal of Behavioral Nutrition and Physical Activity*, 12(106). doi:10.1186/s12966-015-0269-2.

Chapter 1

1. Ofcom (2016) 'Children and parents: Media use and attitudes report.' Available at: https://www.ofcom.org.uk/research-and-data/media-literacy-research/children/children-parents-nov16.
2. 'Screen-based lifestyle harms children's health.' *Guardian*, 25 December 2016. Available at: https://amp.theguardian.com/education/2016/dec/25/screen-based-lifestyle-harms-health-of-children.
3. Ince, D.C., Swearingen, C.J. and Yazici, Y. (2009) 'Wrist pain in 7–12-year-olds playing with game consoles/handhelds: Younger children have more pain, independent from time spent playing.' *Arthritis & Rheumatism*, 60: 1234.
4. Bener, A., Al-Mahdi, H.S., Vachhani, P.J. et al. (2010) 'Do excessive internet use, television viewing and poor lifestyle habits affect low vision in school children?' *Journal of Child Health Care*, 14(4): 375–85. doi:10.1177/1367493510380081.
5. Swing, E.L., Gentile, D.A., Anderson, C.A. et al. (2010) 'Television and video game exposure and the development of attention problems.' *Pediatrics*, 126(2): 214–21. doi:10.1542/peds.2009-1508.
6. Rideout, V. (2014) 'Learning at home: families' educational

media use in America.' A survey by the Joan Ganz Cooney Center. Available at: http://www.joanganzcooneycenter.org/publication/learning-at-home.

7. Weis, R. and Cerankosky, B.C. (2010) 'Effects of video game ownership on young boys' academic and behavioral functioning: a randomized, controlled study.' *Psychological Science*, 21(4): 463–70. doi:10.1177/0956797610362670.

8. Uhls, Y.T., Michikyan, M., Morris, J. et al. (2014) 'Five days at outdoor education camp without screens improves preteen skills with nonverbal emotion cues.' *Computers in Human Behavior*, 39: 387–92. http://doi.org/10.1016/j.chb.2014.05.036.

Chapter 2

1. Ofcom (2016) 'Children and parents: Media use and attitudes report.' Available at: https://www.ofcom.org.uk/research-and-data/media-literacy-research/children/children-parents-nov16.

2. A TLF panel survey conducted on behalf of children's clothing retailer Vertbaudet. Reported in: http://www.express.co.uk/life-style/health/591277/Mobile-phone-tablet-baby-children-technology-Vertbaudet.

3. A survey of 1,500 parents by Internet Matters to mark Safer Internet Day (February 2017). Reported in: http://www.dailymail.co.uk/news/article-4197494/Almost-HALF-six-year-olds-online-bedrooms.html.

4. Liddle, E.B., Hollis, C., Batty, M.J. et al. (2011) 'Task-related default mode network modulation and inhibitory control in ADHD: effects of motivation and methylphenidate.' *Journal of Child Psychology and Psychiatry*, 52(7): 761–71. doi:10.1111/j.1469-7610.2010.02333.x.

Chapter 3

1. Ofcom (2012) 'Children and parents: Media use and attitudes report.' Available at: https://www.ofcom.org.uk/__data/assets/pdf_file/0020/56324/main.pdf.
2. Ao.com survey of 1,000 British mums, September 2015. Available at: http://www.mirror.co.uk/news/technology-science/technology/over-three-quarters-british-mums-6455379.
3. Rideout, V.J., Roehr, U.G. and Roberts, D.F. (2010) 'Generation M²: Media in the lives of 8–18-year-olds.' A Kaiser Family Foundation study. Available at: https://kaiserfamilyfoundation.files.wordpress.com/2013/04/8010.pdf.
4. Gentile, D.A., Choo, H., Liau, A. et al. (2011) 'Pathological video game use among youths: a two-year longitudinal study.' *Pediatrics*, 127(2): e319–29. doi:10.1542/peds.2010-1353.
5. Christakis, D.A., Zimmerman, F.J., DiGiuseppe, D.L. et al. (2004) 'Early television exposure and subsequent attentional problems in children.' *Pediatrics*, 113(4): 708–13.
6. Gentile, D.A., Reimer, R.A., Nathanson, A. et al. (2014) 'Protective effects of parental monitoring of children's media use: a prospective study.' *JAMA Pediatrics*, 168(5): 479–84.

Chapter 4

1. 2015/16 figures from the National Child Measurement Programme. Reported in Mayor, S. (2016) 'Over a third of children aged 10–11 in England are overweight or obese.' *BMJ*, 355: i5948. Available at: http://www.bmj.com/content/355/bmj.i5948.

2. Sebastian Coe writing in the *Daily Telegraph*, 8 April 2014.
3. Griffiths, L.J., Cortina-Borja, M., Sera, F. et al. (2013) 'How active are our children? Findings from the Millennium Cohort Study.' *BMJ Open*, 3:e002893. doi:10.1136/bmjopen-2013-002893.
4. Figures from the British Heart Foundation National Centre for Physical Activity and Health, based at Loughborough University. Available at: http://www.telegraph.co.uk/news/health/news/12108895/tablet-generation-means-nine-in-10-toddlers-live-couch-potato-lives.html.
5. Nightingale, C.M., Rudnicka, A.R., Donin, A.S. et al. (2017) 'Screen time is associated with adiposity and insulin resistance in children.' *Archives of Disease in Childhood*, 13 March. doi:10.1136/archdischild-2016-312016.
6. Study run by the National Center for Education Statistics (2011) 'Kindergarten class of 2010–11 (ECLS-K:2011).' Available at: https://nces.ed.gov/ecls/kindergarten2011.asp.
7. Gentile, D.A., Reimer, R.A., Nathanson, A. et al. (2014) 'Protective effects of parental monitoring of children's media use: a prospective study.' *AMA Pediatrics*, 168(5): 479–84.
8. Aggio, D., Fairclough, S., Knowles, Z. et al. (2012) 'Temporal relationships between screen time and physical activity with cardiorespiratory fitness in English schoolchildren: A 2-year longitudinal study.' Available at: https://www.researchgate.net/publication/224913575_Temporal_relationships_between_screen-time_and_physical_activity_with_ cardiorespiratory_fitness_in_English_Schoolchildren_A_2-year_longitudinal_study.
9. Maddison, R., Ni Mhurchu, C., Jull, A. et al. (2007) 'Energy expended playing video console games: an opportunity to

increase children's physical activity?' *Pediatric Exercise Science*, 19(3): 334–43.

doi:http://dx.doi.org/10.1123/pes.19.3.334.

10. Maddison, R., Foley, L., Ni Mhurchu, C. et al. (2011) 'Effects of active video games on body composition: a randomized controlled trial.' *American Journal of Clinical Nutrition*, 94(1): 156–63. Available at: http://ajcn.nutrition.org/content/94/1/156.short.

11. Mattriciani, L., Olds, T. and Petrov, J. (2012) 'In search of lost sleep: Secular trends in the sleep time of school-age children and adolescents.' *Sleep Medicine Review Journal*, 16(3): 203–11. doi:10.1016/j.smrv.2011.03.00516(3).

12. Hale, L. and Guan, S. (2015) 'Screen time and sleep among school-aged children and adolescents: A systematic literature review.' *Sleep Medicine Review Journal*, 21: 50–58. doi:10.1016/j. smrv.2014.07.007.

13. Survey by Tiger Mobiles conducted by polling agency Carter Digby in the UK in October 2016 among 1,635 adults who have children aged between five and fourteen. Available at: https://www.tigermobiles.com/2016/11/majority-parents-worried-childrens-screen-time-bed-survey-finds.

14. Higuchi, S., Motohashi, Y., Liu, Y. et al. (2005) 'Effects of playing a computer game using a bright display on presleep, physiological variables, sleep latency, slow wave sleep and REM sleep.' *Journal of Sleep Research*, 14(3): 267–73.

15. Dworak, M., Schierly, T., Bruns, T. et al. (2007) 'Impact of singular excessive computer game and television exposure on sleep patterns and memory performance of school-aged children.' *Pediatrics*, 120(5): 978–85. Available at: http:// pediatrics.aappublications.org/content/120/5/978.

Chapter 5

1. Griffiths, M.D. and Hunt, N. (1998) 'Dependence on computer games by adolescence.' *Psychological Reports*, April: 475–80.
2. Weinstein, A.M. (2010) 'Computer and video game addiction – a comparison between game users and non-game users.' *American Journal of Drug & Alcohol Abuse*, 36: 268–76. doi:10.3109/00952990.2010.491879.
3. Han, D.H., Bolo, N., Daniels, M.A. et al. (2011) 'Brain activity and desire for internet video game play.' *Comprehensive Psychiatry*, 52(1): 88–95.
4. Kardaras, N. (2016) 'It's "digital heroin": How screens turn kids into neurotic junkies.' *New York Post*, 27 August. Available at: http://nypost.com/2016/08/27/its-digital-heroin-how-screens-turn-kids-into-psychotic-junkies.
5. Hou, H., Jia, S., Hu, S. et al. (2012) 'Reduced striatal dopamine transporters in people with internet addiction disorder.' *Journal of Biomedicine & Biotechnology*. doi.org/10.1155/2012/854524.
6. Lin, F., Zhou, Y., Du, Y. et al. (2012) 'Abnormal white matter integrity in adolescents with internet addiction disorder: a tract-based spatial statistics study.' *PloS ONE* 7(1). doi.org/10.1371/journal.pone.0030253.

Chapter 6

1. 'More than half of parents unaware of age limit on social media.' Online survey of 4,000 people in the UK, 2,608 of whom were parents. Carried out in November 2016 by

ResearchBods Ltd on behalf of the NSPCC. Available at:
https://www.nspcc.org.uk/what-we-do/news-opinion/social-
media-age-limit.

2. Knowthenet's The Social Age survey (October 2016) canvassed
1,006 parents of children aged eight to sixteen, and 1004
children aged eight to sixteen. Available at:
http://213.248.242.14/articles/kids-not-equipped-coming-
digital-age-nine.

3. 'Measuring national well-being: Insights into children's mental
health and well-being' (2015). Available at: https://www.ons.
gov.uk/peoplepopulationandcommunity/wellbeing/articles/
measuringnationalwellbeing/2015-10-20.

4. Best, P., Manktelow, R. and Taylor, B. (2014) 'Online
communication, social media and adolescent wellbeing: a
systematic narrative review.' *Children and Youth Services Review*.
doi:10.1016/j.childyouth.2014.03.001.

5. A study of 500 teachers conducted by school trips provider
JCA (2010). Reported in: http://www.telegraph.co.uk/
education/educationnews/8142721/Social-networking-teachers-
blame-Facebook-and-Twitter-for-pupils-poor-grades.html.

6. Clark, C., Hawkins, L., National Literacy Trust (2010)
'Young People's Reading: The importance of the home
environment and family support.' Available at: http://www.
literacytrust.org.uk/research/nlt_research/2055_young_
people_s_reading_the_importance_of_the_home_
environment_and_family_support.

7. Children's Commissioner (2017) 'Growing up digital.' Available
at: http://www.childrenscommissioner.gov.uk/publications/
growing-digital.

Chapter 7

1. Childwise press release. Available at: http://www.childwise.co.uk/uploads/3/1/6/5/31656353/childwisc_press_release_-_monitor_2016.pdf.
2. Knowthenet's The Social Age survey. Available at: http://213.248.242.14/articles/kids-not-equipped-coming-digital-age-nine.
3. NSPCC (2016) 'What children are telling us about bullying: Childline report 2015–16.' Available at: https://www.nspcc.org.uk/globalassets/documents/research-reports/what-children-are-telling-us-about-bullying-childline-bullying-report-2015-16.pdf.
4. 'Cyberbullying triples according to new McAfee "2014 Teens and the Screen" study.' Study carried out by McAfee. Available at: https://www.mcafee.com/us/about/news/2014/q2/20140603-01.aspx.
5. Hinduja, S. and Patchin, J.W. (2015) 'Cyberbullying victimization.' Study by the Cyberbullying Research Center. Available at: http://cyberbullying.org/2015-data.

Chapter 8

1. Mazurek, M.O. and Wenstrup, C. (2013) 'Television, video game and social media use among children with ASD and typically developing siblings.' *Journal of Autism and Developmental Disorders*, 43(6): 1258–71. doi:10.1007/s10803-012-1659-9.
2. Mazurek, M.O. and Engelhardt, C.R. (2013) 'Video game use and problem behaviors in boys with autism spectrum

disorders.' *Research in Autism Spectrum Disorders*, 7(2): 316–24. doi.org/10.1016/j.rasd.2012.09.008.

3. Engelhardy, C.R., Mazurek, M.O. and Sohl, K. (2013) 'Media use and sleep among boys with autism spectrum disorder, ADHD, or typical development.' *Pediatrics*, 132(6): 1081–9. doi:10.1542/peds.2013-2066.

4. Cross, E.-J., Richardson, B., Douglas, T. et al. (2009) 'Virtual violence: protecting children from cyberbullying.' Beatbullying, London.

Chapter 9

1. Study conducted by gaming price comparison and swap site, www.Playr2.com, which surveyed 1,221 parents of children aged 17 who frequently played video games. Available at: http://kotaku.com/5901395/two-thirds-of-parents-admit-they-dont-bother-checking-video-game-age-ratings.

2. Anderson, C.A., Shibuya, A., Ihori, N. et al. (2010) 'Violent video game effects on aggression, empathy, and prosocial behavior in Eastern and Western countries: a meta-analytic review.' *Psychological Bulletin*, 136(2): 151–73. doi:10.1037/a0018251.

3. Carnagey, N., Anderson, C. and Bushman, B. (2007) 'The effect of video game violence on physiological desensitization to real-life violence.' *Journal of Experimental Social Psychology*, 43: 489–96. doi:10.1016/j.jesp.2007.04.007.

4. Gentile, D.A., Lynch, P.J., Linder, J.R. and Walsh, D.A. (2004) 'The effects of violent video game habits on adolescent

hostility, aggressive behaviors, and school performance.' *Journal of Adolescence* 27(1): 5–22. doi:10.1016/j.adolescence.2003.10.002.

5. Przybylski, A.K., Deci, E.L., Rigby, C.S. et al. (2014) 'Competence-impeding electronic games and players' aggressive feelings, thoughts, and behaviors.' *Journal of Personality and Social Psychology*, 106(3): 441–57.

6. Adachi, P.J.C. and Willoughby, T. (2011) 'The effect of video game competition and violence on aggressive behavior: which characteristic has the greatest influence?' *Psychology of Violence*, 1(4): 259–74. doi:10.1037/a0024908.

7. Lemmens, J.S., Valkenburg, P.M. and Peter, J. (2011) 'The effects of pathological gaming on aggressive behavior.' *Journal of Youth and Adolescence*, 40(1): 38–47.

8. Ortiz de Gortari, A.B., Aronsson, K. and Griffith, M. (2011) 'Game transfer phenomena in video game playing: a qualitative interview study international journal of cyber behavior, psychology and learning.' *International Journal of Cyber Behavior, Psychology and Learning (IJCBPL)*. doi:10.4018/ijcbpl.2011070102.

Chapter 10

1. Survey for the charity, Action for Children (January 2016). Available at: https://www.actionforchildren.org.uk/news-and-blogs/whats-new/2016/january/unplugging-from-technology.

2. Blum-Ross, A. and Livingstone, S. (2016) 'Families and screen time: current advice and emerging research.' LSE Media Policy Project, Media Policy Brief 17. The London School of

Economics and Political Science, London, UK. Available at: http://eprints.lse.ac.uk/66927/1/Policy%20Brief%2017-%20 Families%20%20Screen%20Time.pdf.

3. Poll of more than 1,000 parents carried out by ComRes for Channel 4 News. Available at: http://www.mirror.co.uk/news/ technology-science/technology/rise-iparenting-survey-reveals-parents-6487093.

Chapter 11

1. 'Making Learning Mobile' project by Project Tomorrow and Kajeet. Available at: http://www.tomorrow.org/publications/ MakingLearningMobile.html.

2. Formby, S. (2014) 'Parents' Perspectives. Children's Use of Technology in the Early Years.' The Literacy Trust.

3. Flewitt, R., Messer, D. and Kucirkova, N. (2914) 'New directions for early literacy in a digital age: The iPad.' *Journal of Early Childhood Literacy*, 15(3): 289–310. doi:10.1177/1468798414533560.

4. Dye, M.W., Green, C.S. and Bavelier, D. (2009) 'The development of attention skills in action video game players.' *Neuropsychologia*, 47(8-9): 1780–9. doi:10.1016/j.neuropsychologia.2009.02.002.

5. Green, C.S. and Bavelier, D. (2003) 'Action video game modifies visual selective attention.' *Nature*, 423: 534–7.

6. Olson, C.K. (2010) 'Children's motivation for video game play in the context of normal development.' *Review of General Psychology*, 14(2): 180–7.

7. Lieberman, D.A., Chamberlin, B., Medina, E. et al. (2011) 'The power of play: Innovations in Getting Active Summit 2011: a

science panel proceedings report from the American Heart Association.' *Circulation*, 123(21): 2507–16. doi:10.1161/CIR.0b013e318219661d.

8. Maskey, M., Lowry, J., Rodgers, J. et al. (2014) 'Reducing specific phobia/fear in young people with autism spectrum disorders (ASDs) through a virtual reality environment intervention.' *PLoS One*. doi.org/10.1371/journal.pone.0100374.

Chapter 12

1. Ofcom (2014) 'Adults' media use and attitudes report.' Available at: https://www.ofcom.org.uk/__data/assets/pdf_file/0020/58223/2014_adults_report.pdf.
2. Survey of 1,500 adults carried out by OnePoll for First Direct. Results reported in: http://www.independent.co.uk/life-style/gadgets-and-tech/news/britons-spend-62m-hours-a-day-on-social-media-thats-an-average-one-hour-for-every-adult-and-child-8567437.html.
3. Study by Opinion Matters for the New Forest National Park Authority. Results reported in: http://www.telegraph.co.uk/technology/news/10981242/Screen-addict-parents-accused-of-hypocrisy-by-their-children.html.
4. Bandura, A., Ross, D. and Ross, S.A. (1961) 'Transmission of aggression through imitation of aggressive models.' *Journal of Abnormal and Social Psychology*, 63: 575–82.
5. Campaign video for Start-rite's 'Pass It On' campaign. Reported in: http://www.dailymail.co.uk/femail/article-3194143/Children-reveal-parents-addiction-mobile-phones-makes-REALLY-feel-thought-provoking-new-video.html.

6. Hiniker, A., Schoenebeck, S.Y. and Kientz, J.A. (2016) 'Not at the dinner table: parents' and children's perspectives on family technology rules.' *CSCW '16*, 1376–89. doi:10.1145/2818048.2819940.

Index